More Praise for HOT T

"In an important new book, H̶e̶ Singer sums up the evidence on global warming as neither settled, nor compelling, nor even very convincing."
   —CHICAGO TRIBUNE

"**HOT TALK, COLD SCIENCE** is a unique and in-depth analysis of an important component of the global warming issue. I would encourage anyone involved with the global warming issue to give this book a serious hearing. Singer is right, this topic is definitely 'unfinished business'."
   —WILLIAM M. GRAY
      Professor of Atmospheric Science,
      Colorado State University

"**HOT TALK, COLD SCIENCE** is an important, comprehensive, timely, and thorough book."
   —WILLIAM A. NIERENBERG
      Director Emeritus, Scripps Institution of Oceanography

"**HOT TALK, COLD SCIENCE** is immensely enjoyable. Dr. Singer examines diverse theoretical and empirical studies and presents the results in an accessible and highly readable book."
   —SALLIE BALIUNAS
      Astrophysicist, Center for Astrophysics,
      Harvard-Smithsonian Observatory

"**HOT TALK, COLD SCIENCE** is of much interest: I've long followed the literature on the long and short term temperature records, ice ages, etc. I like especially the many charts and graphs in the book. One of the refreshing ideas is that the worry about the melting of the icecaps could be the reverse: enough increased evaporation from the oceans to make the caps grow!"
   —H. RICHARD CRANE
      Member, National Academy of Sciences

Hot Talk
Cold Science

Global Warming's Unfinished Debate

Revised Second Edition

S. Fred Singer

Foreword by
Frederick Seitz

The INDEPENDENT INSTITUTE

Oakland, California

The Independent Institute
100 Swan Way, Oakland, CA 94621-1428
Telephone: 510-632-1366 • Fax 510-568-6040
E-mail: info@independent.org
Website: http//www.independent.org

ISBN 0-945999-81-X

Published by The Independent Institute, a nonprofit, nonpartisan, scholarly research and educational organization that sponsors comprehensive studies on the political economy of critical social and economic issues. Nothing herein should be construed as necessarily reflecting the views of the Institute or as an attempt to aid or hinder the passage of any bill before Congress.

Library of Congress Cataloging-in-Publication Data

Singer, S. Fred (Siegfried Fred), 1924–
    Hot Talk, cold science : global warming's unfinished debate / S. Fred Singer ; foreword by Frederic Seitz.
        p.   cm.
    Inludes bibliographical references and index.
    ISBN 0-945999-78-X. – ISBN 0-945999-81-X (pb)
    1. Global warming–Government policy. 2. Climatic changes–Goverement policy. 3. Green-house gases–Government policy.
    I. Title.
    QC981.8.G56S55 1998
    363.738'74–dc21   98-11557
                                                                    CIP

Contents

Acknowledgments

T his book is based on a more detailed, forthcoming volume, *Global Warming: Unfinished Business*, under the auspices of The Independent Institute. I am most grateful to the Institute's president, David Theroux, who first conceived of this book, and whose assistance has been essential throughout in making the book possible. I gratefully acknowledge the research support received from numerous sources including the Atlas Economic Research Foundation, Electric Power Research Institute, Lynne and Harry Bradley Foundation, Smith-Richardson Foundation, Jacobs Family Foundation, and Independent Institute. Research assistance was provided by Sean R. McDonald, editorial assistance by Candace C. Crandall, and general support by Douglas P. Houts. Numerous colleagues, who critically reviewed the manuscript, made important scientific contributions to this book.

Climate change is a very complicated subject, touching on many disciplines. I would be grateful for critical comments from readers and will make every effort to incorporate these in future printings.

Foreword

For scientists wanting fame and fortune, it has become far easier to pander to irrational fears of environmental calamity than to challenge them. But Professor Fred Singer has never been one to lean on conventional wisdom. An atmospheric and space physicist, he has unassailable scientific credentials. This book, *Hot Talk, Cold Science,* will be difficult to dismiss, though many, in their rush to establish international agreements and poorly conceived policies and regulations, will undoubtedly wish to do so.

Fred Singer has been a pioneer in many ways. As an academic scientist in the 1950s, he published the first studies on subatomic particles trapped in the Earth's magnetic field—radiation belts later discovered by physicist James Van Allen. Also, in challenging the findings of other scientists, he was the first to make the correct calculations for using atomic clocks in orbit, hence contributing to the verification by satellites of Einstein's general theory. He further designed satellites and instrumentation for remote sensing of the atmosphere, accomplishments for which he received a White House Presidential Commendation.

Switching careers in the 1960s, he established and served as first director of the U.S. Weather Satellite Service, now part of the National Oceanographic and Atmospheric Administration (NOAA); his efforts were recognized with the U.S. Department of Commerce Gold Medal Award. Dr. Robert M. White, former NOAA Administrator and later President of the National Academy of Engineering, wrote of Singer's achievement: "The contribution that Fred made to the development of the operational weather satellite system was crucial to its successful launch...His understanding of space technology and remote sensing put him in an outstanding position to chart the course of that very important component...some of his fundamental ideas about the use of space vehicles for atmospheric observation were turned into reality."

Returning to university life in the 1970s, Fred Singer's concern with the environment led him to investigate the effects of human activities on the atmosphere. In 1971, he calculated that population growth (together with increased rice growing and cattle raising) would cause a substantial upward trend of methane, an important greenhouse gas that could contribute to climate warming. He also predicted that methane, once it reached the stratosphere, would be transformed into water vapor, leading to a possible depletion of stratospheric ozone. The fact that methane levels are indeed rising was discovered a few years later; that stratospheric water vapor is also increasing was finally demonstrated in 1995.

At the core of Fred Singer's arguments on the global warming issue is a desire to more fully understand the mechanisms that cause climate to change—in response to natural or manmade forcing—and, perhaps more important, to secure a place for science outside the realm of selfish bureaucracy or the reach of irrational environmentalism.

It is one thing to impose drastic measures and harsh economic penalties when an environmental problem is clear-cut and severe. It is quite another to do so when the environmental problem is largely hypothetical and not substantiated by careful observations. This is definitely the case with global warming. As Professor Singer demonstrates—and his views are backed by many in the scientific profession, including myself—we do not at present have convincing evidence of any significant climate change from other than natural causes.

Until we do, it would be a reckless breach of trust to put in force hasty policies that create real personal and economic hardships for most of the world's population.

FREDERICK SEITZ
President Emeritus, Rockefeller University
Past President, National Academy of Sciences

Preface

The United States and other industrialized nations are on the brink of adopting policies that will ruin national economies, and drive manufacturing and other industries into less developed and less regulated countries (with the perverse effect of destroying their environments). Such policies will cost citizens literally hundreds of billions of dollars in higher product costs and lost wages—all to mitigate climate "disasters" that exist only on computer printouts and in the feverish imagination of professional environmental zealots.

Why? The proposed actions to mandate legally binding emissions targets for carbon dioxide will curtail energy use through rationing and taxation; the impact of these actions will fall most heavily on the poorest among us. On July 25, 1997, the United States Senate, very sensibly, voted 95-0 against adopting such policies if they would damage the U.S. economy or if they were not uniformly applied to all nations. But the Senators, very unsensibly, failed to address the more basic issue—the lack of scientific justification for such policies. It is still not too late to do so.

U.S. Undersecretary of State Timothy Wirth has made the statement—repeated by bureaucrats around the world—that climate science is "settled." President Bill Clinton, in calling for sweeping policy actions, has termed the evidence for global warming "compelling."

The purpose of this book is to demonstrate that the evidence is neither settled, nor compelling, nor even convincing. On the contrary, scientists continue to discover new mechanisms for climate change and to put forth new theories to try to account for the fact that global temperature is not rising, even though greenhouse theory says it should.

Significantly, the United Nations' own science advisory group, the Intergovernmental Panel on Climate Change, has repeatedly backtracked in trying to explain the disparity between computer simulations of the atmosphere and actual observations. As late as 1992, the IPCC judged the data and the theory to be "broadly consistent" and claimed a "scientific consensus" that global warming was upon us. In its 1996 report, however, the IPCC had to admit that the models were unreliable; it brought out a new explanation—atmospheric aerosols—to paper over the gap between theory and observations.

Now, only a year later, this IPCC hypothesis no longer works. As this book points out, there are some half-dozen plausible mechanisms that could account for the fact that—despite computer predictions of a major warming trend—no significant global warming has been observed in the last half-century, and none at all in the last two decades.

No one knows which of these mechanisms, if any, is correct; discovering that is the job of the research scientist. *All that we can be certain of at this point is that the science of global warming is truly "unfinished business."*

September 1997

Preface to second edition

A number of recent developments have made a second edition desirable:

- Research underlying climate change has uncovered new facts, largely invalidating the evidence on which the 1996 IPCC conclusions were based. Far from being "settled," as frequently claimed, climate science has moved ahead rapidly on all fronts and is uncovering new problems. (See page 73.)
- The economic impact of a possible global warming has been re-evaluated and found to yield positive benefits rather than losses. This new analysis, if sustained, undercuts the need for drastic mitigation policies, or even the mandatory emission cuts called for by the Kyoto Protocol. (See page 25.)
- The Kyoto Protocol of December 1997, even if punctiliously observed, is shown to have negligible effects on any future greenhouse warming. It would require the United States to reduce carbon dioxide emission from energy-fuel use by 7% (below 1990 levels), or by about 35% around the year 2010. Kyoto is not only costly but also ineffective. (See page 68.)

This second edition also incorporates a number of corrections, mainly of figure labels. I am grateful to the many alert readers who have called them to my attention.

April 1999

S. FRED SINGER
Fairfax, Virginia

1

Overview

The Scientific Case Against the Global Climate Treaty

A driving force behind the push for a global climate treaty has been the United Nations' Intergovernmental Panel on Climate Change (IPCC). Through a series of well-publicized reports—co-authored by teams of scientists and policy specialists— the IPCC has come to be viewed by many governmental agencies, environmental policy organizations and the media as the leading source of scientific information on climate change. It is for this reason that I focus much of my attention on reports issued by this esteemed organization.

The major conclusion of the United Nations' Intergovernmental Panel on Climate Change (IPCC WG-I 1996)—that "the balance of evidence suggests a discernible human influence on global climate" cannot and should not be used to validate current Global Circulation Models (GCMs). The growing discrepancy between weather satellite observations, backed by balloon radiosonde data, and the results of computer models, throws doubt on the models' adequacy to predict a future warming. An earlier IPCC (1990) conclusion that observed and calculated temperature changes are "broadly consistent" is no longer accepted; the current IPCC explanation of the acknowledged discrepancy in terms of cooling effects from anthropogenic (man-made) sulfate aerosols is being increasingly disputed. There exist different, competing views about the cause(s) of the discrepancy—including exogenous factors like solar variability, and endogenous factors like clouds or water vapor distribution—all inadequately treated by current computer models. The models do not include a variety of human influences, ranging from possible climate effects of air traffic to the diversion of fresh water from the Mediterranean.

1

Even if a moderate warming were to materialize, its consequences would be largely benign—for other climate parameters, for sea-level changes, and for agricultural production. The goal of the Global Climate Treaty—avoiding a "dangerous" level of greenhouse (GH) gases—cannot as yet be scientifically defined; higher GH gas levels may well produce a more stable climate. Therefore, the prudent course is to practice a "no-regrets" policy of conservation and efficiency improvements and rely on adaptation to meet any damaging effects of climate change. At the same time, building on successful initial experiments, the capability of ocean fertilization to draw down atmospheric CO_2 should be demonstrated.

The chief points of this overview are as follows:

1. The major conclusion of the 1996 report of the U.N.-sponsored science advisory group, the Intergovernmental Panel on Climate Change (IPCC), is that "the balance of evidence suggests a discernible human influence on global climate" (IPCC WG-I 1996, Chapter 8). This innocuous but ambiguous phrase has been (mis)interpreted to mean that computer models predicting a future warming have now been validated. But such a connection is specifically denied in the body of the IPCC report (IPCC WG-I 1996, p. 434)—although not in the politically approved IPCC Summary for Policymakers (SPM).

On the contrary, the global temperature record of this century, which shows both warming and cooling, can best be explained by natural climate fluctuations caused by the complex interaction between atmosphere and oceans, and perhaps stimulated by variations of solar radiation that drives the Earth's climate system. The satellite record of global temperature, spanning nearly twenty years, does not show a global warming—much less one of the magnitude predicted by General Circulation Models (GCMs). The gap between the satellite observations and existing theory is so large that it throws serious doubt on all computer-modeled predictions of future warming. Yet this discrepancy is never mentioned in the IPCC report's Summary—nor does the SPM even admit the existence of satellites (Singer 1996).

If one were to extrapolate the maximum allowed temperature trend from satellites to the year 2100, allowing for a further increase in atmospheric carbon dioxide and other greenhouse gases (GHG), one might estimate the increase in global average temperature at about 0.5°C (U.S. General Accounting Office 1995)—about one-fourth of the

Box 1. The Global Climate Treaty and its Implementation

Under the terms of the Global Climate Treaty (officially, the U.N. Framework Convention on Climate Change—FCCC), efforts are currently underway to establish a protocol for reducing emissions of greenhouse gases (GHG), especially carbon dioxide (CO_2) from the burning of fossil fuels. The announced objective of the FCCC (in Article 2) is to "achieve stabilization of greenhouse gas concentrations in the atmosphere at a level that would prevent *dangerous* anthropogenic interference with the climate system" (emphasis added). Such an emission control scheme, with legally binding targets and timetables, would be extremely costly and have a detrimental impact on the world economy. It would reduce by several percent the GDPs of industrialized countries, responsible for most of the world's energy consumption, and countries that export fossil fuels will experience a loss of income (Ismail 1997). But developing countries will suffer as well, since their well-being and economic stability depend on world trade and general prosperity. (Montgomery 1997; Goldemberg 1995)

These major economic losses, initially borne by agricultural and industrial producers, will be passed on to consumers in the form of higher prices. The loss of jobs, together with a higher cost of living, will cause extreme hardship, especially for the poor. Such economic sacrifices cannot be justified by any conceivable environmental benefits. *The Climate Treaty, signed at the Earth Summit in Rio de Janeiro in June 1992, rests on three suppositions that are questionable or even demonstrably false:*

1. It assumes that a global warming signal has been detected in the climate record of the last hundred years, thus validating the predictions of a major future warming by computer models (IPCC WG-I 1996).

2. It futher assumes that a substantial warming in the future will produce catastrophic consequences, including droughts, floods, storms, a rapid and significant rise in sea level, a collapse of agriculture, and a spread of tropical diseases.

3. Finally, it assumes that we know which atmospheric concentrations of greenhouse gases are "dangerous" and which are not; that drastic reductions of emissions of carbon dioxide—and energy use—by industrialized nations can indeed stabilize CO_2 concentration at near-present levels; and that such economically damaging measures can be justified politically—even though there is no significant scientific support for a global "threat" of climate warming.

"best" IPCC value—hardly detectable and completely inconsequential.

Any future warming would be reduced further by the cooling effects of volcanoes—a factor not specifically considered by the IPCC. Even though we cannot predict the exact dates of future volcanic eruptions, we have sufficient statistical information about past eruptions to estimate an average cooling effect.

2. Even if global warming were to occur, it would most likely lead to positive benefits overall rather than to disbenefits. Human activities, especially agriculture, have always thrived during warm periods and faltered during cold periods (Moore 1995). A greenhouse warming should lead to a reduction in severe storms. Furthermore, it seems likely that a global warming will lower, rather than raise, sea levels, because more evaporation from the ocean would increase precipitation and thereby thicken the ice caps of Greenland and Antarctica (Singer 1997a).

3. Finally, no credible attempt has been made to define what constitutes a "dangerous" level of atmospheric CO_2; thus the goal of the U.N. Climate Treaty (the Framework Convention on Climate Change-FCCC) is arbitrary. If one chooses as the target the present concentration, about 30 percent higher than the preindustrial CO_2 level of 280 parts per million by volume (ppmv), emission rates must be cut by over 60 percent on a worldwide basis, according to IPCC modeling (IPCC WG-I 1990, p. xi).

A policy of adaptation to a possible climate change should be considered rather than energy rationing. If it becomes advisable to limit the growth of atmospheric CO_2, it might be more cost effective to speed up CO_2 absorption into the ocean, rather than by reducing emissions. In fact, by fertilizing the ocean with micronutrients, it may be possible to increase phytoplankton and fish populations, and thereby derive commercial benefits from the excess atmospheric CO_2 (Cooper et al. 1996).

A discussion of the underlying science will reveal the following:

1. There Is No Detectable Anthropogenic Global Warming

The fear has often been expressed that anthropogenic release of GH gases could cause a temperature increase larger and more rapid than anything experienced in human history. The geologic record gives us an important perspective on this issue.

Carbon Dioxide: The paleo record of atmospheric carbon dioxide

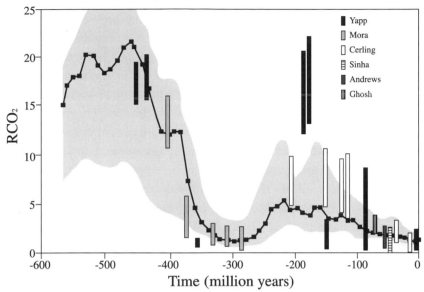

Fig. 1. A compilation of carbon dioxide concentrations (as multiples of the pre-industrial value of 280 ppmv) [Berner 1997, reprinted with permission, © American Association for the Advancement of Science]. In spite of considerable uncertainties, one can discern a rapid downward trend starting about 450 Myr ago, and an ongoing downward trend that began about 200 Myr ago. The recent anthropogenic increase would be difficult to notice.

(Berner 1997, Fig. 1) shows considerable changes. The CO_2 concentration was about twenty times the preindustrial value 500 million years (Myr) ago, diminished rapidly as CO_2 was removed by weathering and reached its lowest level about 300 Myr ago. The concentration of CO_2 then rose to about 4 to 5 times present levels and has been steadily diminishing ever since—with considerable fluctuations. Concerns have been raised that too low a CO_2 level would be catastrophic for plant growth (Idso 1989).

Temperature: Climate has always varied. The paleotemperature records from ocean sediments and from Greenland and Antarctic ice cores have established the existence of 17 ice-age (glacial/interglacial) cycles in the past 2 million years. We are now in the Holocene, the interglacial warm period that began approximately 11,000 years ago, ending the most recent Ice Age.

Historical records document the existence of a "Little Ice Age," a period of colder than average global temperatures, between about 1450 and 1850, and the "Medieval Climate Optimum" around 1000 A.D.

Even larger and more rapid temperature variations, certainly not caused by human activities, can be found in high-resolution ocean-core data (Fig. 2), going back 3,000 years (Keigwin 1996). Abrupt climate transitions occurred simultaneously at the equator and both polar regions around 8,000 years ago (Stager and Mayewski 1997), giving credence that such changes were worldwide. These findings contradict the IPCC claim that a human-caused greenhouse warming would lead to temperature changes that are greater and/or faster than any experienced up until now.

The Climate Treaty Goal: The climate record gives little guidance as to what constitutes atmospheric levels of CO_2 producing "dangerous interference with the climate system"; the announced purpose of the Climate Treaty is to avoid such levels. We do not know, however, whether climate variability depends on CO_2 concentration. Rapid variations in temperature appeared during the most recent ice age when the atmospheric CO_2 concentration was only about 200 parts per million (ppm), rather than 280 ppm, the preindustrial value during most of the Holocene (IPCC 1996). Stager and Mayewski (1997) suggest that the warmer Holocene was "relatively more stable than the late Pleistocene." If this observation is supported by further data, we should be striving to increase, not decrease, the CO_2 level in the atmosphere.

Azar and Rodhe (1997) placed the "dangerous" CO_2 level in the range of 350-400 ppmv, and warned that a 2° C increase would be "dangerous." Their assumption that the present level of CO_2 would have

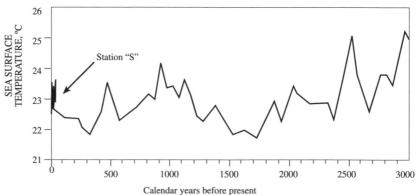

Fig. 2. **Detailed temperature variations of the past 3,000 years (during recorded history), as determined from ocean sediment studies in the North Atlantic. Note the rapid variations, as well as much warmer temperatures 1,000 and 2,500 years ago [Keigwin 1996, reprinted with permission, © American Association for the Advancement of Science].**

an appreciable probability of producing such a large temperature increase is dubious.

Consequently, the goal of the Treaty remains scientifically undefined. While the 1990 IPCC report favored CO_2 stabilization at the 1990 level of about 350 ppm, the 1996 report appears to aim for a politically more acceptable level of 550 ppm, double the preindustrial value. *But without scientific guidance, the goal is entirely arbitrary, and any stabilization level—or none—will do.*

Natural climate variations: The cause of climate variations is largely unknown, and therefore unpredictable. Many climate scientists believe in the existence of irregular, quasi-periodic oscillations based on purely internal interactions between atmosphere and ocean, which computer models cannot yet simulate. Other scientists hold that solar variability is the main cause of climate variations. Indeed, striking correlations have been observed between sunspot cycles and climate (Fig. 3) (Lassen and Friis-Christensen 1991). Unfortunately, we do not understand how the rather small variations in solar radiation can influence climate so dramatically—although there are now a number of different mechanisms offered as explanations.

Whatever the cause, the GCMs used to predict future climate increases are not able to account for natural fluctuations of climate. Even the most sophisticated GCMs, using coupled atmospheric and ocean computer models, are unable to reproduce the El Niño-South-

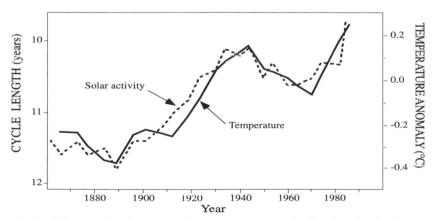

Fig. 3. Solar activity, in terms of sunspot cycle length (broken line), shows a strong correlation with global temperatures (solid line) [Friis-Christensen and Lassen 1991, reprinted with permission, © American Association for the Advancement of Science]. The authors extended the correlation back to 1500 by using proxy data [Lassen and Friis-Christensen 1995].

ern Oscillation (ENSO), the North Atlantic Oscillation (NAO), or other quasi-periodic variations of climate.

Human influences?: The existence of natural climate variability makes it difficult to detect any "signal" that may be due to human activities. In the 1996 IPCC WG-I report, the authors of Chapter 8 claim that they can discern human influence gradually emerging from the "noise" of natural climate fluctuations as the concentration of CO_2 increased with time. A correlation coefficient between observed and computed geographic climate patterns that appears to increase with time (Fig. 4) was shown by B.D. Santer et al. (1995; IPCC WG-I 1996, Chapter 8). In a contemporaneous research publication, however, some of the same authors express a different opinion (Barnett et al. 1996) (see Box 2).

There are several additional problems with the IPCC claim of a "discernible human influence":

The "natural" variability is derived from computer runs of GCMs, rather than from actual observations, and therefore likely to be different from the true value (see Box 2).

Box 2. Is There a Discernible Human Influence in Global Climate? Two Views by the Same Researchers:

From IPCC WG-I 1996, Summary of Chapter 8, p. 412, by B.D. Santer, T.M.L. Wigley, T.P. Barnett, E. Anyamba:

"...there is evidence of an emerging pattern of climate response to forcings by greenhouse gases and sulphate aerosols... from the geographical, seasonal and vertical patterns of temperature change...These results point towards a human influence on global climate."

From T.P. Barnett, B.D. Santer, P.D. Jones, R.S. Bradley and K.R. Briffa, The Holocene, 6, 255-265, 1996:

"Estimates of...natural variability are critical to the problem of detecting an anthropogenic signal...We have estimated the spectrum...from palaeo-temperature proxies and compared it with... general circulation models.... None of the three estimates of the natural variability spectrum agree with each other... Until... resolved, it will be hard to say, with confidence, that an anthropogenic climate signal has or has not been detected."

The computed climate pattern (IPCC WG-I 1996) includes, of course, the effects of the rise in greenhouse gases, but only one of

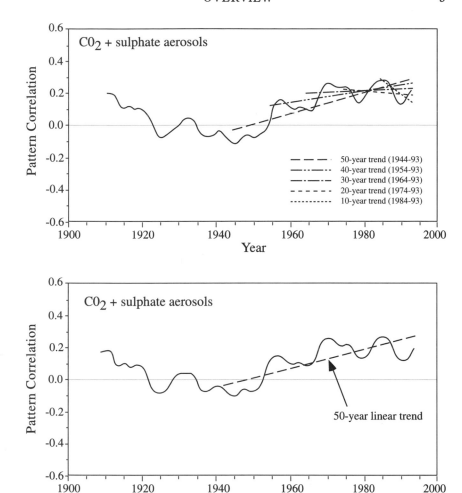

Fig. 4. a) [Santer et al. 1995, reprinted with permission, © Springer-Verlag]. As can be seen, the correlation coefficient between calculated and observed temperature patterns is rather small and shows strong variations. It decreases during the period of rapid warming (before 1940) and does not increase during the past 25 years when atmospheric CO_2 level rose greatly. b) In the 1996 IPCC report (Fig. 8.10b) only the increasing 50-year trend line is shown; the zero and negative trend lines were omitted by the authors for reasons that were not explained.

the cooling effects of particulates—the direct albedo effects of human-produced sulfate aerosols. Their indirect effects leading to cloud production—important but difficult to quantify—are omitted, as are the radiative effects of mineral dust and of smoke and soot from biomass burning; yet these may be the most important effects (Fig. 5)

(Schwartz and Andreae 1996).

Explaining the discrepancy: "Climate sensitivity" is defined as the temperature rise calculated by GCMs for a doubling of CO_2-equivalent GH gases; the IPCC reports (1990, 1996) quote values between 1.5°C and 4.5°C. The clearest demonstration that current GCMs are inadequate comes from a comparison between their "best" predicted warming of 0.3°C per decade (IPCC 1990, 1996) and the actual observations. Surface measurements with thermometers show a warming of 0.13°C per decade since 1979, while global satellite measurements using a microwave sensor actually show a slight cooling of the lower troposphere—about -0.04°C per decade (Fig. 6) (Spencer and Christy 1992; Christy 1997). Thus there are two separate problems:

a. The difference between surface and satellite measurements.

b. The difference between observations and computer model results. There are several possible causes, each with different implications:

i. External man-made causes (aerosols, ozone changes).

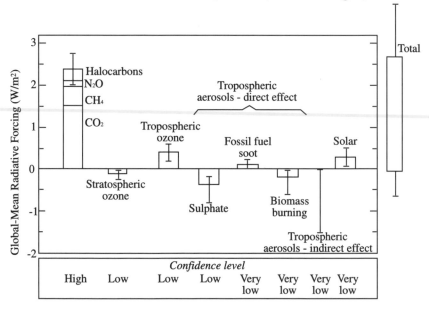

Fig. 5. Global-mean radiative forcing for greenhouse gases and particulates. Note the huge uncertainties shown by the vertical lines, especially for the indirect effects of aerosols. The total forcing could even be negative, but there is little reality to a "global mean" because of the large geographic and temporal changes of particulates [Schwartz and Andreae 1996, reprinted with permission, © American Association for the Advancement of Science].

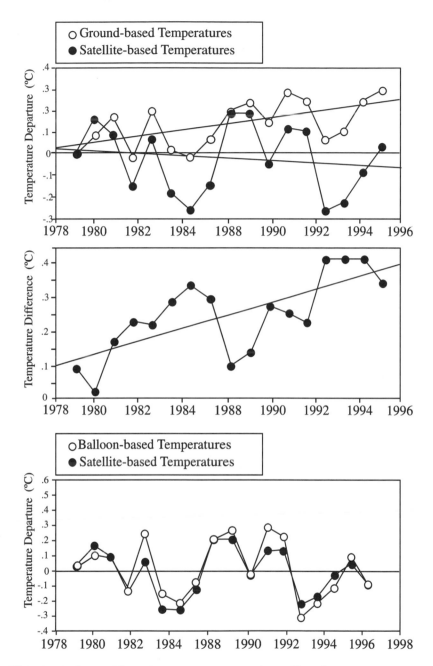

**Fig. 6. a) Ground-based temperatures and satellite-observed tempera-
tures show quite different trends. Surface data shows a slight positive
temperature trend while satellite data shows a cooling trend. b) The difference between the two data sets is large, statistically significant, and
growing. c) The independent balloon radiosonde data support the satellite data [World Climate Report 1996].**

ii. External natural causes (solar variations, volcanoes).

iii. Internal (to the models) causes (clouds, water vapor distribution).

a. Satellite vs. Surface Data: The IPCC report and individual investigators (Hansen et al. 1995; Hurrell and Trenberth 1996) have attempted to account for the discrepancy between surface and satellite data by claiming that they are both correct but measure different quantities. This explanation might cover a short period of time, but becomes untenable when the discrepancy extends over many years and keeps growing. The temperatures of the lower troposphere and surface cannot move apart for very long.

More recently, Hurrell and Trenberth (1997) claim to have found an error created when different satellite records are joined together. They assert that the reported cooling trend is an artifact. Against this assertion is the fact that the balloon record of the lower troposphere agrees almost exactly with the satellite record, but not with the trend in the surface record. Furthermore, the surface observations are subject to an "urban heat island" (UHI) effect (Fig. 7) (Goodridge 1996). As population grows in the vicinity of weather stations, an artificial warming trend is introduced that is difficult to identify and eliminate.

The urban heat island effect has also been demonstrated in other areas where there is a dense network of weather stations. After correcting for the UHI, the years around 1940 emerge as the warmest years of the century both in the U.S. record (Karl and Jones 1989) and the European record (Balling 1997) (see Fig. 8).

b. Climate Observations vs. Computer Results: The discrepancy between observations and computer model results is very serious. Even if the satellite data are corrected for the effects of El Niño and volcanic eruptions (Christy and McNider 1994), the growing discrepancy in trends indicates that the GCMs are not adequately simulating atmospheric processes.

i. Attempts have been made to "fix" the discrepancy by introducing into the GCMs the effects of sulfate aerosols said to cool and counteract the positive radiative forcing of greenhouse gases (Taylor and Penner 1994; Mitchell et al. 1995). But a recent modeling experiment indicates that the aerosol effect is minute and ascribes the lack of troposphere warming to a cooling of the stratosphere, presumably caused by an ongoing depletion of stratospheric ozone (Tett et al.

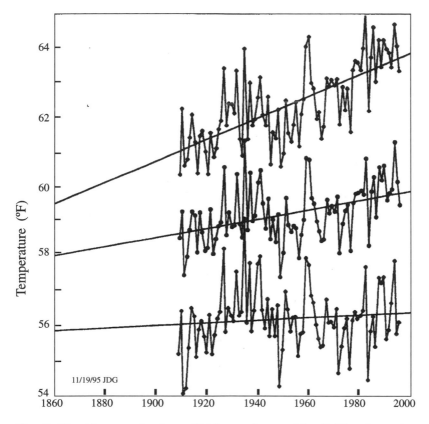

Fig. 7. Warming trends since 1910, as observed by California stations. The top curve is for counties with populations of more than 1 million, the middle curve is for populations between 0.1 and 1 million, and the bottom curve for counties with less than 0.1 million inhabitants. Note the increased warming trend for populous counties, indicating an urban heat island effect [Goodridge 1996].

1996). Comprehensive analysis by Hansen et al. (1997) concludes that radiative forcing by aerosols is a minor factor in climate modeling—contrary to the main thrust of the Second Assessment Report of the IPCC (1996).

An effort to reconcile observations of the northern hemisphere (NH) and southern hemisphere (SH) also shows that aerosols cannot account for the discrepancy between computer model results and observation. Man-induced sulfate aerosols are restricted almost entirely (~90 percent) to the NH and thus should have opposed or slowed GH warming mainly in the NH. Yet recent reanalyses of SH land data (which are the most credible and also predicted by models to change

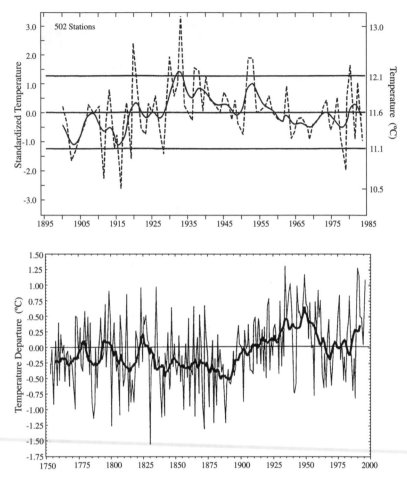

Fig. 8. After correcting for the urban heat island effect (UHI), the years around 1940 emerge as the warmest years of the century in both the US record (top) [Karl and Jones 1989] and European record (bottom) [Balling 1997].

more than sea surface temperatures) have cut the observed rate of warming of the SH to about half of the NH (Jones 1994; Hughes and Balling 1996).

A fairly conclusive piece of evidence against the importance of aerosols comes from the observation of a strong (zonal) warming trend in northern midlatitudes. It can be seen in surface, balloon and satellite data (Fig. 9). Its cause is suspected to be a positive radiative forcing from cirrus clouds produced by the rapidly increasing air traffic in the vicinity of the tropopause (Singer 1997b). It is therefore surprising that efforts are still underway to explain the discrepancy

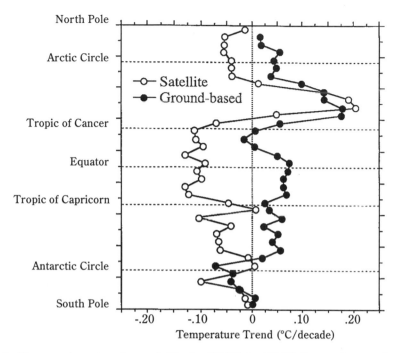

Fig. 9. Temperature trends vs. latitude (1979 to 1995) for surface and satellite data. A fairly conclusive piece of evidence against the importance of aerosols comes from the observation of a strong (zonal) warming trend in northern mid-latitudes [Michaels 1996, p. 7] . Aerosols which are supposed to cool the climate, originate mainly in the Northern Hemisphere at middle latitudes.

between observations and climate models in terms of the direct effects of aerosols (Hasselmann 1997; Mitchell and Johns 1997). The confident IPCC (1996) conclusion that "...the balance of evidence suggests a discernible human influence on global climate," seems no longer supportable (Kerr, 1997) (see Box 3).

ii. Solar variability in the ultraviolet region of the spectrum (Lean 1991) and through solar wind could also modulate the Earth's climate. One possibility is the climate influence of the well-known 11-year cycle of stratospheric ozone thickness (Haigh 1996). Another possibility is that the solar-modulated 11-year cycle of cosmic-ray intensity influences climate (Dickinson 1975; Markson and Muir 1980). Several mechanisms have been discussed: a. the generation of nitrogen oxides, depleting stratospheric ozone (Bauer 1982); b. ionization by cosmic rays that affect the rate of freezing of supercooled water in high-level clouds (Tinsley and Deen 1991); c. changes in the vertical air-earth current density affecting atmospheric dynamics (Tinsley 1996); d. changes in cloud cover correlated with cosmic-ray intensity (Svensmark and Friis-Christensen 1997) (see Fig. 10).

iii. It is likely, however, that the discrepancy is due to inadequately treated internal effects within the models rather than external factors, like changes in aerosols or ozone. Possible internal effects include a negative feedback from increased cloudiness or from a possible reduction of upper troposphere water vapor—either by an intensification of the Hadley circulation (Ellsaesser 1984, 1990) or by meso-

---

Box 3. Doubts about the IPCC Conclusions

Richard Kerr (1997):

"Many scientists say it will be a decade before computer models can confidently link the warming to human activities."

Benjamin Santer, as quoted by Kerr (1997):

"We say quite clearly that few scientists would say that the attribution issue was a done deal." "It's unfortunate that many people read the media hype before they read the (IPCC) chapter (on the detection of global warming)".

Brian Farrell, as quoted by Kerr (1997):

"There really isn't a persuasive case being made for detection of greenhouse warming...the error bars are as big as the signal."

Klaus Hasselmann (1997) concludes that:

"...uncertainties in the detection of anthropogenic climate change can be expected to subside only gradually in the next few years while the predicted signal is still slowly emerging from the natural climate variability noise."

Thomas Crowley, wishfully, in the *New York Times* (2 July 1997):

"...statistical studies suggest that we are already on the verge of detecting a greenhouse warming."

---

scale drying mechanisms related to cloud dynamics (Lindzen 1990). It is too early to tell which of the many feedback possibilities can account for the shortcomings of the current GCMs, although satellite infrared (IR) and microwave observations may settle the issue (Spencer and Braswell 1997). But until validated by actual climate observations, the model results should not be used as a basis for developing

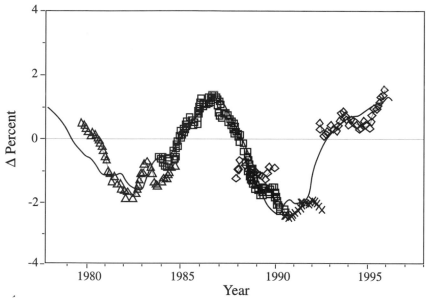

Fig. 10. Changes in cosmic-ray intensity at Climax, Colorado (thick line) and cloud data sets from four satellites (triangles are Nimbus 7 data, squares are ISCCP.C2 data, diamonds are DMSP data, crosses are ISCCP.D2 data) [Svensmark and Friis-Christensen 1997]. The data are smoothed using a 12-months running mean.

policy. Contrary to the confident assertions by politicians that climate science is "settled," climate science is still "unfinished business."

It is important to understand the reason for the discrepancy: if due to aerosols, the ongoing growth of GH gases will eventually win out, due to the short atmospheric lifetime of aerosols (a few days) compared to the lifetime of CO_2. If the discrepancy is due to solar variations, then the enhanced GH effect should again win out, because solar variations are presumably periodic and average out to zero over the long term. However, if the discrepancy is due to internal effects, leading to reduced or even negative feedbacks, then climate sensitivity itself can be greatly reduced and any future warming is likely to be unimportant.

2. Historically, the Consequences of Modest Warming Are Positive

We already know from recorded human history that warm periods are beneficial for human populations and that cold periods bring disaster in the form of crop failure and disease (Moore 1995). In fact, weather should improve as a result of global warming. The IPCC re-

port mentions that North Atlantic hurricanes diminished in both frequency and severity in the past 50 years (IPCC WG-I 1996, figure 3.19,
p. 170). If greenhouse warming serves to reduce the equator-to-pole
temperature gradient, as predicted, then midlatitude storms should
diminish in intensity as well.

The most feared consequence of global warming has been the
possibility of a catastrophic sea-level rise (IPCC WG-I 1996). It is vir-
tually impossible to predict (purely from theory) whether sea level
will rise or fall as climate warms. On the one hand, melting glaciers
and thermal expansion of ocean water will lead naturally to a rise in
sea level (Wigley and Raper 1992). On the other hand, increased evaporation from the oceans and subsequent precipitation and accumulation of ice on Greenland and especially Antarctica would lower sea
level (Oerlemans 1982). The only way to settle this issue is by exami-
nation of data. Fig. 11 shows that both global average temperature

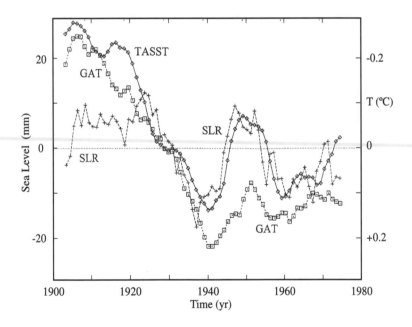

**Fig. 11. Sea-level rise change (SLR) (after subtracting the linear rise),
global average temperature (GAT), and tropical average sea-surface temperature (TASST) (Bottomley et al. 1990). Note that the correlation is
inverse, suggesting that global warming could slow down, rather than
speed up sea-level rise [Singer 1997a]. (All plots are 10-year running
means of yearly values.)**

(GAT) and tropical average sea-surface temperature (TASST) are anti-correlated with fluctuations of sea level rise (SLR). There appears to be an anti-correlation also between the (rather limited) data on ice accumulation and SLR. Furthermore, the major El Niño events of this century correlate with negative peaks in sea level. Results from GCM calculations also suggest that polar ice accumulation will outweigh thermal expansion and glacier melting (Thompson and Pollard 1995). *All of these findings point to the conclusion that future warming will slow down rather than accelerate the ongoing rise in sea levels* (Singer 1997a).

As far as agriculture is concerned, a modest global warming based on increased CO_2 levels—as seems likely over the next 100 to 200 years—is bound to be beneficial (Idso 1995). The predicted rise in average temperature comes mainly from a rise of night-time and winter temperatures, i.e. a reduction in the diurnal and seasonal temperature range (Karl et al. 1991). Thus, greenhouse warming should lead to fewer frosts and longer growing seasons. In addition, the increased CO_2 will stimulate plant growth, and at the same time reduce the need for water (Idso 1989; Wittwer 1992). A "greening" of the Earth at northern high latitudes has already been reported (Myneni et al. 1997), as well as an earlier spring growing season (Keeling, Chin and Whorf 1996). In any case, economists have shown empirically that farmers are quite capable of adjusting to climate changes (Mendelsohn et al. 1994).

Another concern, that warmer temperatures would spread insect-borne tropical diseases (Patz et al. 1996) neglects the fact that insect control and public health are the primary determinants of such diseases, and that, with cheap and rapid mass transportation, the human vector is increasingly important. Frequent epidemics of malaria and yellow fever occurred in the United States and even Northern Russia when the climate was much colder (U.S. Department of Agriculture 1952), and wealthy Singapore, situated at the equator, does not share the widespread tropical diseases prevalent in poverty-stricken Africa.

3. Control of Atmospheric CO_2 through Ocean Fertilization: An Alternative to Emission Controls *

With the residence time of atmospheric CO_2 variously estimated as between 50 and 200 years, its current excess over its preindustrial value will eventually be absorbed by biota on land and in the ocean. But even if a future warming is negligibly small and on the whole beneficial, there may still be political pressure to control the level of atmospheric carbon dioxide. The standard approach, and the one most appealing to international negotiators, has been to control CO_2 emission rates; for example, the 1990 IPCC report pointed out that maintaining the current level of CO_2 (350 ppm) would require a worldwide emission reduction of more than 60 percent from 1990 rates (IPCC WG-I 1990)—with a corresponding reduction in energy use. Stabilizing at the 550 ppm level, approximately double the preindustrial value, requires an emission reduction of some 50 percent. It is hard to imagine broad political support for such a plan, given its disastrous economic consequences. Fully realizing this, politicians have instead advocated more modest reductions of between 5 and 20 percent from current rates, with more to come later. *Even if these reduced rates were achieved on a worldwide basis, however, they would only slow down—at great cost—the current upward trend of atmospheric CO_2. Stabilizing emissions does not stabilize concentration if the atmospheric residence time is long enough so that CO_2 accumulates.*

An alternative approach to emission control is to sequester the CO_2 from the atmosphere—or at least demonstrate that sequestering is technically and economically feasible. The conventional approach to CO_2 sequestration has called for tree planting on a massive scale, thereby tying up CO_2 for decades, to be released when the wood decays. But tree planting can be costly and impractical; it requires huge areas and great expenditures of funds to make an appreciable impact. Order-of-magnitude figures for sequestration by trees are 0.8-1.6 tons of carbon per hectare per year (Nordhaus 1991a, 1991b); thus, to absorb current production of CO_2 requires about 50 million km^2 (ca. 4,500 x 4,500 miles!).[1]

An equivalent, but economically far more attractive approach is to speed up the natural absorption of CO_2 into the ocean. Currently, much of the world's oceans are a biological desert. Even though many

---

* For a more complete discussion of mitigation, see the Appendix.

of these areas have adequate supplies of the basic nutrients, nitrates and phosphates, they lack essential micronutrients like iron. The biomass of phytoplankton in the world's oceans amounts to only one to two percent of the total global plant carbon; yet these organisms fix between 30 to 50 billion metric tons of carbon (GtC) annually, which is about 40 percent of the total. (For reference, the atmosphere contains 750 GtC in the form of CO_2.)

Ocean fertilization (McElroy 1983) has been widely discussed among scientific specialists, with experiments proposed by the late John Martin (1994), and endorsed by the late Professor Roger Revelle, director of the Scripps Oceanographic Institution in La Jolla, California.[2] With the completion and publication of the successful IronEx-II test (see papers in the Oct. 10, 1996 issue of *Nature*), it now makes sense to consider ocean fertilization as a candidate technology for sequestering atmospheric CO_2.

Building on the scientific success of IronEx-II, a large-scale demonstration could prove the technical and economic feasibility for lowering the content of atmospheric CO_2 at a fraction of the cost now contemplated for emissions reduction. While it may never be necessary to reduce atmospheric CO_2, it would be comforting to know that we have the technical capability to do so.

An additional benefit of fertilization is that it converts the current excess of atmospheric CO_2 into an important resource, to be exploited for feeding a growing world population. Large-scale fertilization of areas of the Pacific and the Southern Oceans for the purpose of stimulating the growth of phytoplankton would draw down atmospheric CO_2 without depressing the economies of industrialized nations or limiting the economic growth options of developing nations. With phytoplankton as the base of the oceanic food chain, any increase in that population can lead to the development of new commercial fisheries in areas currently devoid of fish. *Carbon dioxide from fossil fuel burning thus becomes a natural resource for humanity rather than an imagined menace to global climate.*

4. Adjusting to Climate Change

It is reasonably certain that any effects of human-induced climate change will be minor also compared to other sources of change over the next century. Climate is important mainly because of its effect on natural resources (e.g., water, land, plants, forests, habitats, and other biological resources) and on human activities, such as agriculture, forestry, human settlements, and recreation, which depend on natural resources. Based upon existing assessments, human-induced climate change over the next hundred years will be much less important to the environment than the other agents of global change, i.e., population growth, economic growth, and technological changes.

Also, existing assessments tend to overestimate negative impacts of climate change, while underestimating positive ones (see Box 4).

It is a fundamental principle of public policy that problems which are the most important and can be reduced, if not eliminated, at the least cost to society should be given the highest priority. Accordingly, one must address the question: How important is a possible climate change—above and apart from the major variations of natural origin—compared with other agents of future global change? It is difficult to justify major expenditures to address climate change in the presence of other unmet societal needs, e.g., improved public health, food security, education, and personal and public safety. *If, as argued above, climate change is a minor problem, then adaptation to climate change becomes the preferred option; any resources saved can be directed to more important societal problems.* (Goklany 1992, 1995)

Adaptation to climate change is, of course, the normal response to seasonal and interannual variations of climate, and to many extreme climate events as well. Adaptation is generally easier for technologically advanced societies and for societies with resources, which can afford adequate housing, heating, air-conditioning, etc. Throughout human history populations have adapted successfully to large permanent climate changes; for example, when Germanic tribes migrated from the frozen north to the Mediterranean.

While adaptation to climate change may be problematic for natural ecosystems, the ability to adapt is, paradoxically, highest for those economic sectors and human activities most sensitive to climate change. Because of their sensitivity to climate, such systems have

Box 4. Some Benefits of Increased CO_2 Levels

It should be noted that little, if any, of the now over $2 billion per year environmental research budget has been used to identify, document, or quantify possible benefits of adding CO_2 to the atmosphere, or of any of the other consequences of man's activities. This bias in itself has contributed greatly to public perception that these activities pose serious threats. That there are benefits from adding CO_2 to the atmosphere is undeniable:

1. *Fertilization of the biosphere:* CO_2 is essential to plant life. At the last glacial maximum (18,000 years BP) the CO_2 level dropped to ~190 ppmv, which is close to the level where plants would begin to experience propagation failure.

2. *Longer frost-free growing season:* Any GH warming will be due to reduced radiative cooling of the surface, thus will be greatest in winter, at higher latitudes and at night. That is, minimum temperatures will be affected much more than maximum temperatures, leading to longer frost-free growing seasons and little, if any, additional summer-afternoon heat stress.

3. *Greater water efficiency for plants:* Except in already arid subtropics, precipitation is predicted to increase. Increased CO_2 allows plants to ingest the CO_2 they need with less opening of their stomata, thus making it possible to survive with less water since less is lost by evapotranspiration.

4. *Health:* Increase of minimum temperatures with little effect on maximum temperatures will be beneficial both in reducing cold stress on health and in reducing the requirements for space heating. Idso (1989) has suggested that "the significant downturn in circulatory heart disease experienced worldwide over the past two decades" is a possible consequence of the 25 percent increase in CO_2 in the atmosphere. Respiration is controlled by the concentration of CO_2 (rather than of oxygen) in the blood. Thus CO_2-stimulated deeper breathing may have reduced the strain on the circulatory system.

always been heavily managed and have long histories of successful and rapid adoption of technological and management innovation.

Besides energy conservation and the encouragement of nonfossil-fuel resources, actions meeting adaptation and development goals include increasing the productivity or efficiency of crops, livestock, forests, fisheries, and human settlements, consistent with principles of sustainable development.

Adaptation to climate change is one of the most desirable policy options; ignoring adaptation overestimates the negative impacts of climate change. In contrast, strategies such as control of CO_2 emission from fossil-fuel combustion may compromise society's ability to cope with other global problems that require economic development. The most serious climate threat to mankind may be a return of an ice age following the end of the current warm interglacial period. (see Box 5)

Response strategies and impact assessment reports by the IPCC and other groups point out that developing countries are more vulnerable to climate change, not because climate change is expected to be greater in those nations—climate change will be least in the tropical zone—but because of lack of financial and technical resources. Hence, it is imperative to expand the level of these resources. This can be done through sustainable economic growth and technological change, which will reduce poverty and eventually population growth rates. These, in turn, require the establishment of the appropriate legal, eco-

---

Box 5. Additional Reasons for Avoiding Hasty Actions.

1. We have not been able to define scientifically the goal of the Climate Treaty, namely the GH gas concentrations considered as "nondangerous." A higher CO_2 level may turn out to be preferable to a lower one.

2. Postponing action by 10 years or so will not appreciably affect future temperatures (Schlesinger and Jiang 1991).

3. A number of economic arguments have been advanced as to why it pays to postpone action on limiting emissions (Wigley, Richels and Edmonds 1996), without compromising the ability to act later. For instance, capital equipment can be replaced at lower cost after it wears out. In the interim, new information about underlying climate science will enable us to make more rational decisions.

nomic, and institutional framework to encourage more economic growth and technological change.

In summary, successful adaptation to climate change requires specific actions—many of which will also help limit greenhouse gas emissions—that will stimulate sustainable economic growth and continued technological progress. Meeting these twin goals is critical to ensuring that limitation of greenhouse gases, if it should become necessary, would cause the least disruption to society.

Economic Benefits from Global Warming: A Post-IPCC Re-Evaluation

A team of economists has re-evaluated the impact of a modest warming on the U.S. economy (Mendelsohn and Neumann 1999). The 1996 IPCC report had assembled five published estimates of damages, ranging from $55 to $139 billion per year (in 1990 dollars, including market and non-market sectors). These estimates, published between 1991 and 1995, differ greatly in details — some assigning the greatest loss to agriculture, others to flooding from sea-level rise.

The new work derives a *net benefit* to the U.S. economy, rather than a net loss. In addition to including certain benefits that had been overlooked in prior studies, the new methodology considers the possibility of adaptation, as well as the beneficial effects of increased CO_2 on agriculture. See Table 1 below.

If these improved estimates are borne out, then it will no longer be possible to carry out a cost-benefit analysis for the mitigation of greenhouse warming. For if the net benefits of warming are indeed positive (adding also the appreciable benefits of a reduction in sea-level rise), then one should do nothing to oppose such a warming. Indeed, the only policy that makes sense under those circumstances is the same "no-regrets" policy that is called for if the science does not support the warming predictions of climate models: as much conservation as makes economic sense, and higher efficiencies in the use of energy and other resources.

Table 1 corresponds to Summary Table 12-2 from Mendelsohn and Neumann (1999). "Previous Estimate" refers to the five estimates published in the IPCC Report, Working Group III, p. 203 [Nordhaus 1991, Cline 1992, Fankhauser 1995, Tol 1995, and Titus 1992], and show annual (market) losses ranging from \$14 to 68 billion. The "New Estimate" are the analyses assembled by Mendelsohn and Neumann (1999), showing an overall gain to the U.S. economy from warming. Note: Nonmarket effects include all recreation sector impacts as well as most nonconsumptive components of the water resources estimates (excluding hydroelectric production, which is included in the market effects estimate).

Table 1. Estimated Annual Impact of Effective Doubling of CO_2
(Billions in 1990 U.S. Dollars)

Market and Non-Market Sector — Methodological Improvements	New Estimate: +2.5° C, + 7% precipitation		Previous Estimate
	2060 Economy	1990 Economy	1990 Economy
Market Sector Impact Estimates:			
Agriculture — Inclusion of additional crops and adaptation opportunities:			
	+\$41.4	+\$11.3	-\$1 to -\$18
Timber — Dynamic climate, ecological, and timber modeling:			
	+\$3.4	+\$3.4	\$-1 to -\$44
Water Resources (Market Only) — Integrated hydrologic and economic models:			
	-\$3.7	-\$3.7	-\$7 to -\$16
Energy — Includes all space conditioning fuels:			
	-\$4.1	-\$2.5	-\$1 to -\$10
Coastal Structures — Dynamic analysis of representative sites:			
	-\$0.1	-\$0.1	-\$6 to -\$12
Commercial Fishing — First estimate:			
	-\$0.4 to +\$0.4	-\$0.4 to +\$0.4	N/A
Total (Market Sectors):	+\$36.9 +0.2% of GDP	+\$8.4 +0.2% of GDP	-\$14 to -\$68 -0.3% to -1.2% of GDP
Nonmarket Sector Impact Estimates:			
Water Quality — Basin-based regional estimates:			
	-\$5.7	-\$5.7	-\$32.6
Recreation — Includes summer activities and empirical evidence:			
	+\$3.5	+\$4.2	-\$1.7

2

Unfinished Business

Scientific Issues to Be Resolved

A Perspective on a Century of Climate Concerns

C oncern that a man-made increase in greenhouse gases (GHG) causes climate warming goes back to the 19th century. The first definitive paper calculating a rise in temperature was published by the Swedish chemist Svante Arrhenius (1896). Even earlier, the French mathematical physicist J. Fourier showed that certain minor gases, like carbon dioxide (CO_2), in the atmosphere could absorb infrared (heat) radiation emanating from the Earth's surface, interfere with its escape, and thus raise the temperature of the surface appreciably. He compared the effect to that of a greenhouse (GH) (see Fig. 12).

An informative article about the early thinking on greenhouse warming (Weart 1997) reviews the publications that put the subject on the map. In 1938, the British meteorologist G. S. Callendar revived the largely ignored GH hypothesis of Arrhenius. Callendar asserted that the temperature rise since the 1890s was due to greenhouse warming by CO_2. His views were dismissed by his contemporaries. A prominent textbook by T. A. Blair, *Climatology: General and Regional* (Prentice-Hall, 1942), states, "we can say with confidence that climate is not influenced by the activities of man, except locally and transiently." Even as late as 1955, experts argued that the ocean would take up all of the CO_2 entering the atmosphere from fossil-fuel burning, thus invalidating Callendar's arguments.

A more influential objection was that water vapor would dominate greenhouse warming by covering the same spectral region as the carbon dioxide absorption bands; therefore, CO_2 would not add to the greenhouse effect. The American Meteorological Society's 1951

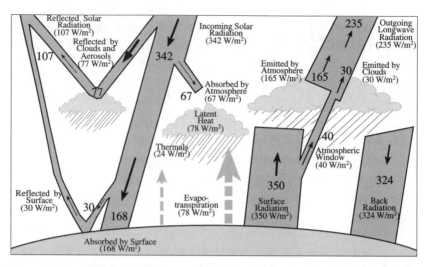

Fig. 12. Earth's radiation energy balance, which controls the way the greenhouse effect works, can be seen graphically here. The numbers in parentheses represent energy as a percentage of the average solar constant-about 342 Wm² —at the top of the atmosphere. Nearly half the incoming solar radiation penetrates clouds and greenhouse gases to the Earth's surface. These gases and clouds reradiate most (i.e., 168 units) of absorbed energy back down toward the surface. This is the mechanism of the greenhouse effect [Schneider 1992].

Compendium of Meteorology, edited by Thomas F. Malone, stated that the carbon-dioxide theory was never widely accepted and "was abandoned when it was found that all of the long-wave radiation (that might be) absorbed by CO_2 is (already) absorbed by water vapor." In 1955, physicist Gilbert Plass, at Johns Hopkins University, pointed out that the carbon dioxide absorption lines did not coincide with water vapor lines once the pressure-broadening effect of the lower troposphere was removed. The Compendium may have been correct, however, in stating that the recent (before 1945) global temperature rise was not related to human activities, but of natural origin.

At about the same time, Hans Suess (Revelle and Suess 1957), at the U.S. Geological Survey, became interested in the ocean uptake of CO_2. But it was Roger Revelle, then director of the Scripps Institution of Oceanography, who was motivated to start long-term measurements of atmospheric CO_2 and, together with Charles David Keeling, discovered that about half of the CO_2 from fossil-fuel burning was remaining in the atmosphere.

Revelle, the "father of greenhouse warming," always regarded the CO_2 increase as a "grand geophysical experiment" which would reveal the consequences of human intervention with the atmosphere. In a paper delivered at a 1975 conference of the American Academy of Arts and Sciences, Revelle pointed to the beneficial effects of CO_2 on agriculture and speculated that the improvements in yield of this century might be connected to a CO_2 rise (Revelle 1977).

In a 1988 interview with *Omni* magazine, Revelle expressed similar optimism—and no great alarm about the risk of global warming. In his July 1988 letters replying to climate concerns by his congressman and then-Senator Timothy Wirth, Revelle advised against drawing any conclusions about global warming from the 1988 drought and warned against taking hasty action. In a co-authored article in the journal of the Cosmos Club shortly before his death in July 1991, Revelle went even further, stating (Singer, Revelle, Starr 1991): *"The scientific base for a warming is too uncertain to justify drastic action at this time."*

Leading up to the Global Climate Treaty

Greenhouse warming has been with us for four and a half billion years, throughout the history of the earth. An earth without infrared-absorbing gases in its atmosphere would be quite cold. We can calculate the surface temperature by balancing the incoming solar radiation with the outgoing heat radiation from the surface. If we assume that the earth (with atmosphere) has an "albedo" of 0.3, i.e., reflects 30 percent of the incoming solar radiation back out into space, and that the infrared (IR) emissivity of the surface is 1.00, the equilibrium temperature would be -18°C, well below the freezing point of water. The average temperature now is 15°C. The difference, about 33°C, can be ascribed to the natural greenhouse effect, produced mainly by the water vapor (WV) and CO_2 of the atmosphere; WV is responsible for well over 95 percent of the GH effect. Without the presence of naturally occurring atmospheric greenhouse gases like carbon dioxide and especially water vapor, our planet would have been a frozen wasteland with oceans covered by ice.[3]

While it has been known for decades that the concentration of CO_2 was rising because of the burning of fossil fuels—oil, gas and

especially coal—precise measurements on a global scale were started
only during the International Geophysical Year in 1957 by Charles
Keeling and the late Prof. Roger Revelle. Their data showed large
seasonal fluctuations, but also a steady upward trend that seemed to
match world consumption of energy, with roughly half of the CO_2 released remaining in the atmosphere (Fig. 13).[4]

One reason for the lack of concern before the 1980s was the fact
that global temperatures had been dropping steadily since about 1940
(Fig. 14). In the early 1970s, an increasing number of climate scientists as well as popularizers were becoming concerned about this downward trend; those of a catastrophic bent saw the temperature decrease
as an indicator of a returning ice age.[5] The literature of the time is
filled with technical and popular papers expounding on such themes.

The climate situation changed drastically around 1975, but only
realized a decade later. Temperatures started to rise rapidly until about
1980, after which, according to satellite data, temperatures stabilized
at about the 1980 level (Fig. 15). We still do not know for certain what

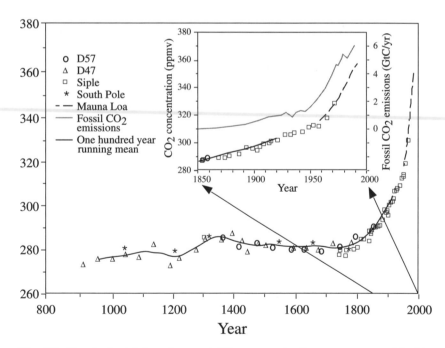

**Fig. 13. Pre-industrial and recent CO_2 concentration as measured in ice
cores and at Mauna Loa, Hawaii. Also shown is the estimated emission
from the burning of fossil fuels. [Adapted from IPCC WG-I 1996]**

happened in the mid-seventies; many suspect that a sudden climate transition took place, involving the ocean circulation.[6]

By 1980, temperatures had nearly reached the 1940 level and the fear of cooling was replaced by a fear of warming, overlooking the fact that warming is generally better for agriculture and most other human activities than is cooling. Budding environmental movements were confronted with the fact that air pollution and water pollution levels were decreasing in Western nations, and new problems had to be found to maintain the enthusiasm and funding for the growing organizations. Global disasters filled the bill, and so we had successive scares on acid rain, ozone depletion, and now on greenhouse (GH) warming (Box 6).

But the subject of climate warming by anthropogenic GH gases came to public notice only as recently as 1988 when a hot summer and major drought devastated much of the agricultural harvest in the United States. In testimony before a Senate committee chaired by then-Senator Albert Gore, NASA climate scientist James Hansen announced that he was "99 percent" sure that global warming was here. From then on, events moved rapidly to the signing of a global climate treaty,

---

Box 6. The Rise of Environmentalism

It is difficult to state exactly when global concerns first arose. Many would place the beginning in the late 1960s, after the appearance of Rachel Carson's *Silent Spring* and Paul Ehrlich's *The Population Bomb*. Certainly, passage of the National Environmental Policy Act and establishment of the U.S. Environmental Protection Agency in 1970 advanced the cause. The first celebration of Earth Day showed that the movement was well underway. Landmark events were the 1972 environmental conference in Stockholm, which launched the acid rain scare, and the publication of the Club of Rome's *Limits to Growth* (Meadows et al. 1972). This apocalyptic work came at a time when public concern about the environment was at its peak. The more sober and scientifically accurate book, *Inadvertent Climate Modification* (SMIC 1971), had relatively little popular impact. On the other hand, the 1971 controversy about the supersonic transport (SST) and the fear of skin cancer from stratospheric ozone depletion aroused great excitement and further escalated global environmental concern.

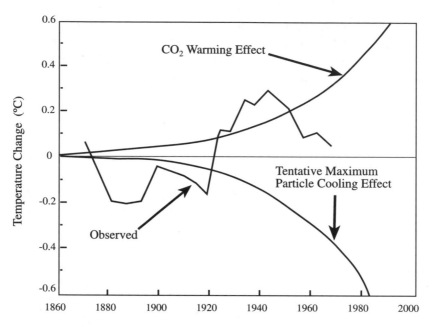

Fig. 14. The general view in the early 1970s was that either warming or cooling may take place [Mitchell 1970; 1975, reprinted with permission, © Kluwer Academic Pub.]. The upper smooth curve represents the estimated warming contribution from increasing carbon dioxide [Manabe and Wetherald 1967]. The lower smooth curve represents the probable maximum cooling contribution by atmospheric aerosol particles [Rasool and Schneider 1971, 1972]. The broken curve refers to the Northern Hemisphere temperature, corrected by Mitchell for the urban heat island effect.

the U.N. Framework Convention on Climate Change (FCCC), in Rio de Janeiro in 1992. Attempts to establish a "protocol" to limit and roll back the rate of emission of GH gases, especially CO_2, followed the treaty. The paradigm was the 1987 Montreal Protocol, which only eight years later led to a ban on the production of ozone-depleting chemicals like chlorofluorocarbons (CFCs). Earlier, in 1988, U.N. agencies had authorized establishment of the Intergovernmental Panel on Climate Change (IPCC) as a scientific advisory body to the FCCC, following the model of the Ozone Trends Panel that paved the way for the Montreal Treaty.

Concurrently, the countries that had ratified the Treaty, convened as a Conference of the Parties (COP), first in Berlin in 1995, then in Geneva in July 1996, and Kyoto in December 1997. COP-1 produced the Berlin Mandate, instructing the Parties to prepare a protocol for

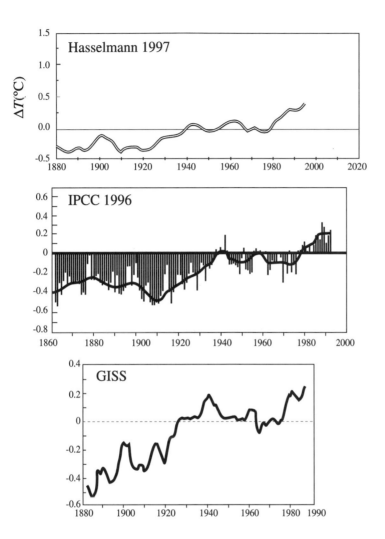

Fig. 15. Global temperature versus time, as determined from surface measurements with thermometers. The temperature changes are referred to an arbitrary baseline. Three different compilations are shown: Hasselmann [1997, reprinted with permission, © American Association for the Advancement of Science]; IPCC [1996], and GISS [Hansen and Lebedeff 1987]. While all three records show the remarkable warming before 1940, likely a natural recovery from the cooling of the "Little Ice Age," the records differ considerably after 1940. The differences between these records illustrate some of the uncertainties in the "global" climate record, based on the selection and treatment of the data.

implementing the Treaty. In the absence of an agreement, COP-2 resulted in a Ministerial Declaration that interpreted the IPCC science report as establishing a firm basis for a future major warming. While the report (IPCC WG-I 1996) only hinted at a "discernible human influence" on climate, the assembled statesmen chose to regard the science as "settled" and proceeded to plan for COP-3 (in Kyoto, December 1997), where countries would present firm proposals for mandatory controls on the emission of greenhouse gases.

Where Climate Policy Stands Now

The ongoing negotiations on the U.N. Climate Convention take for granted that the science of global warming is "settled." Nothing could be further from the truth. The climate models that predict a major warming in the next century have not been validated by observations and therefore cannot—and should not—be used as a basis for decision-making. In spite of the constant use of the phrase "scientific consensus," there is substantial disagreement on many issues within the community of atmospheric scientists and climate specialists (Kerr 1997b; Singer 1997c). Indeed, most scientists believe that the global warming issue should be considered "unfinished business" requiring much further research.

The IPCC produced its first scientific assessment in 1990; it was updated in 1992 in time for the Rio de Janeiro meeting. Its most recent assessment, *Climate Change 1995: The Science of Climate Change*, in May 1996 (IPCC WG-I 1996), arrived at twin conclusions: that the climate changes of the last century are "unlikely to be due entirely to natural fluctuations" and that "the balance of evidence suggests a discernible human influence on global climate."[7] These phrases, while appearing cautious and unobjectionable, can be—and have been— (mis)used to persuade decision-makers that their worst fears about a coming climate catastrophe are backed by a scientific consensus.

Indeed, the second Conference of Parties (COP-2) to the U.N. Framework Convention on Climate Change (FCCC), meeting in Geneva in July 1996, accepted as a basis for urgent policy action the IPCC's main conclusion about a "discernible human influence on climate." The head of the U.S. delegation, Under Secretary of State (for

Global Affairs) Timothy Wirth, proposed legally-binding targets and time frames for emissions of greenhouse gases. He stated: "The science calls upon us to take urgent action." When a number of delegations did not accept this proposal, a Ministerial Declaration by the U.S. and like-minded nations was issued on July 18, calling for a protocol to control emissions of carbon dioxide (CO_2)—and, in effect, to limit the generation of energy. Such global controls on energy use would have serious economic consequences, harming mainly the world's poor, who need low-cost energy for development, for building infrastructure, and for moving from poverty to a higher standard of living.[8]

The policy proposals of the U.S. State Department were presented to a wide audience in a briefing on January 17, 1997. The proposals envisage national emission "budgets"—which are really quotas, or a certain fraction of the global emission permitted for a particular year. This rationing scheme for the use of fossil fuels is supplemented by an emission-trading scheme, which allows any Party to buy unused emission rights from any other Party. In principle, this results in a less-costly realization than strict rationing without trading. The problem arises, of course, in deciding upon the emission budget to be assigned to each nation, and whether to use population or per-capita consumption as a criterion, and whether to use current values or some future value for population and per-capita consumption (see Appendix).

The U.N. Framework Convention on Climate Change (FCCC) states that "…policies and measures to deal with climate change should be cost-effective so as to insure global benefits at the lowest cost" (Article 3.3). The FCCC calls for an "economic system that would lead to sustainable economic growth and development" (Article 3.5). Article 2 states the objective, namely the "stabilization of GHG concentrations…that would prevent dangerous anthropogenic interference with the climate system." *The problem is that we cannot as yet define what that level is, whether the present level, or some future level, or the preindustrial level. Defining dangerous levels is a daunting task for climate science, and should be tackled before far-reaching policies are put in place.[9] Hasty action is ill-advised if the goal is not known.*

General Scientific Controversies

What are the issues on which there is general agreement between the IPCC and its critics, and what are the scientific issues that are in controversy?

There is little dispute about the rising levels of greenhouse gases (GHG), aside from disagreements about sampling and other instrumental questions. There is considerable uncertainty, however, about the future levels of GHG, especially CO_2, since they depend on uncertain assumptions about the growth of population and energy consumption—as well as about the residence time of the excess atmospheric CO_2.

There is important disagreement about the temperature record since 1979, with satellite data showing a slight cooling and surface thermometers showing a warming (Fig. 6). Atmospheric data taken with balloon-borne radiosondes agree with the satellite data. All three sets of observations show much lower trends than what computer models predict. *The key question is why rising levels of greenhouse gases in the atmosphere are not causing a global warming in accord with the expectations from current climate models. Only if these General Circulation Models (GCMs) are validated through observations can one put reliance on their forecasts of future warming.*

The IPCC Summary for Policymakers (SPM) does not confirm that the models have been validated. It merely hints at this possibility by employing the phrase, "The balance of evidence suggests a discernible human influence on global climate." This poorly defined and essentially meaningless conclusion is acceptable only if identified with well-known examples of climate trends, such as the observed stratospheric cooling, a reduction of the diurnal temperature range, decline in the frequency of hurricanes, etc. The IPCC Summary does not, however, validate the GCMs; it does not establish a confirmation of the "climate sensitivity" (deduced from model calculations), given by IPCC as a 1.5 to 4.5°C temperature increase for a doubling of the (equivalent) CO_2 concentration.

GHG concentrations have already gone halfway towards a CO_2-equivalent doubling, mainly in the past 50 years. But the climate record shows no commensurate warming since 1940. The IPCC attempts to explain the disagreement between the temperature record and GH theory by invoking a cooling from man-made aerosols that cancels some of the warming. The realization that the aerosol mechanism cannot bridge the difference between theory and observation and that current climate models are deficient and cannot be relied on to predict future warming will be examined later in this volume.

Nevertheless, some modest future warming should not be ruled out. Its impact will, on the whole, be beneficial for agriculture and other human activities. Even the much-feared sea-level rise may turn out to be a nonproblem.

This critique of the IPCC's conclusions does not argue against prudent policies that conserve energy, increase energy efficiency, or introduce nonfossil energy sources (including nuclear reactors) as alternatives to fossil fuels—as long as such policies make economic sense and are carried out on a voluntary basis.

Scientific Consensus and Peer Review

Much has been made of peer review in validating "good science" as against "fringe science" or "junk science." Peer review is a discretionary tool used by editors of scientific journals. It permits them to screen submitted articles for publication by using anonymous and presumably unbiased referees. Peer review has some value, but it is not perfect; in spite of conscientious referees, many articles are published that contain incorrect material and require later corrections. Also, many new ideas are rejected because they are unfamiliar to the editor(s) and to the referees, or they challenge their deeply held views. Yet science progresses largely because of new ideas or new data that question established wisdom.

The IPCC chapters were never "peer-reviewed" in the generally accepted sense, in spite of such IPCC claims. The IPCC core group prepared the summaries and also selected the reviewers, while in a normal peer review a journal editor will send a scientific article to anonymous referees. There is no record available as to what comments from reviewers were ignored; nor is there a record of minority opinions.

The IPCC report also claims to be based on papers that have been published in the scientific literature and therefore peer-reviewed in the normal sense. However, an analysis by University of Virginia professor Robert Davis shows that some 40 percent of the citations in the 1996 report refer to papers that are "in press," or "to be submitted." For example, the crucial Chapter 8 of the 1996 report deals with "detection and attribution of climate change" and backs up the IPCC conclusion about a "discernible human influence on global climate." The chapter's conclusion is largely based on two papers whose senior author is also the chapter's convening lead author. These two papers appeared well after the IPCC report was accepted; in fact, one appeared only after the IPCC report had already been printed. Chapter 8 contains 18 references to these two unpublished papers; eight of the co-authors of these papers are also listed as contributors to Chapter 8.

The IPCC claims a scientific consensus for global warming by counting not only the 80 or so chapter lead authors who did the actual writing of the most recent Second Assessment Report (SAR) (IPCC WG-I 1996), but also the several hundred contributors who permitted their work to be quoted. The IPCC further includes the hundreds of reviewers, regardless as to whether they agreed or disagreed, or whether their comments were used or not used (Singer 1997c).

The foreword to the 1990 IPCC report admits to the existence of minority views that the editors were "not able to accommodate." There is no indication of the size of the minority or the seriousness of their disagreements. But the extent of consensus can be judged from actual surveys of climate scientists. In 1991, the Science & Environmental Policy Project (SEPP), an independent, foundation-funded research group, mailed questionnaires to 126 U.S. atmospheric scientists. Most of these had contributed to or reviewed the 1990 IPCC report, which has been widely described as presenting a "scientific consensus" about the reality and danger of enhanced greenhouse warming. Colleagues who worked on the report had complained that its "Policymakers Summary" did not accurately represent the conclusions in the report itself. And journalists and bureaucrats presumably read only the Summary, not the 400-page technical report.

The survey results were remarkable. Of over 50 scientists who responded, half agreed that the Summary did not represent the report fairly and could be misleading to nonscientists. An overwhelming majority of respondents agreed that there was no clear evidence in the climate record of the last 100 years for enhanced greenhouse warming due to human activities. About 90 percent agreed with the following statement (on page 254 of the 1990 IPCC WG-I report): "It is not possible to attribute all, or even a large part, of the observed global-mean warming to the enhanced greenhouse effect on the basis of observational data currently available." Only 15 percent believed that current GCMs accurately portrayed the atmosphere-ocean system, and less than 10 percent thought that current GCMs had been adequately validated by the climate record.

Other independent surveys support these findings. For example, a November 1991 Gallup poll of 400 members of the American Meteorological Society (AMS) and the American Geophysical Union (AGU) (actively involved in global climate research) posed the question: "Do you think that global average temperatures have increased during the past 100 years and, if so, is the warming within the range of natural, nonhuman-induced fluctuations?" The poll found that only 19 percent believed that human-induced global warming has occurred.

Also in 1991, Greenpeace International surveyed 400 scientists who had worked on the IPCC report or had published on relevant issues during 1991. Asked whether business-as-usual policies might instigate a runaway greenhouse effect at some (unspecified) future time, only 13 percent of the 113 respondents thought it "probable" and 32 percent "possible." But 47 percent said "probably not"—far from a consensus. This result appears consistent with the results of the SEPP survey. Yet Greenpeace described its survey result as revealing "an as yet poorly expressed fear among a growing number of climate scientists that global warming could lead not just to severe problems but complete ecological collapse."

In 1992, SEPP went a step further and contacted some 300 atmospheric physicists and meteorologists (most of them serving on technical committees of the American Meteorological Society) and asked them to endorse publicly a strongly worded statement (see Box 7) expressing concern that policy initiatives being developed for the 1992 United Nations Earth Summit were being driven by "highly uncertain scientific theories." As a result, more than fifty put their names to the statement.[10]

In July 1996, SEPP released the Leipzig Declaration, based on a conference held in that city in November 1995, cosponsored by the

Box 7. Statement by Atmospheric Scientists on Greenhouse Warming

WASHINGTON, D.C., 27 FEBRUARY 1992—As independent scientists, researching atmospheric and climate problems, we are concerned by the agenda for UNCED, the United Nations Conference on Environment and Development, being developed by environmental activist groups and certain political leaders. This so-called Earth Summit is scheduled to convene in Brazil in June 1992 and aims to impose a system of global environmental regulations, including onerous taxes on energy fuels, on the population of the United States and other industrialized nations.

Such policy initiatives derive from highly uncertain scientific theories. They are based on the unsupported assumption that catastrophic global warming follows from the burning of fossil fuels and requires immediate action. We do not agree.

A survey of U.S. atmospheric scientists, conducted in the summer of 1991, confirms that there is no consensus about the cause of the slight warming observed during the past century. A recently published research paper even suggests that sunspot variability, rather than a rise in greenhouse gases, is responsible for the global temperature increases and decreases recorded since about 1880.

Furthermore, the majority of scientific participants in the survey agreed that the theoretical climate models used to predict a future warming cannot be relied upon and are not validated by the existing climate record. Yet all predictions are based on such theoretical models.

Finally, agriculturists generally agree that any increase in carbon dioxide levels from fossil fuel burning has beneficial effects on most crops and on world food supply.

We are disturbed that activists, anxious to stop energy and economic growth, are pushing ahead with drastic policies without taking notice of recent changes in the underlying science. We fear that the rush to impose global regulations will have catastrophic impacts on the world economy, on jobs, standards of living, and health care, with the most severe consequences falling upon developing countries and the poor.

European Academy for Environmental Affairs. Nearly 100 climate specialists in Europe and U.S.A. affixed their signatures to the document, which was critical of the IPCC conclusions.[11]

Box 8. The Leipzig Declaration
on Global Climate Change

As scientists, we—along with our fellow citizens—are intensely interested in the possibility that human activities may affect the global climate; indeed, land clearing and urban growth have been changing local climates for centuries. Historically, climate has always been a factor in human affairs—with warmer periods, such as the medieval "climate optimum," playing an important role in economic expansion and in the welfare of nations that depend primarily on agriculture. For these reasons we must always remain sensitive to activities that could affect future climate.

Attention has recently been focused on the increasing emission of "greenhouse" gases into the atmosphere. International discussions by political leaders are currently underway that could constrain energy use and mandate reductions in carbon dioxide emissions from the burning of fossil fuels. Although we understand the motivation to eliminate what are perceived to be the driving forces behind a potential climate change, we believe this approach may be dangerously simplistic. Based on the evidence available to us, we cannot subscribe to the so-called "scientific consensus" that envisages climate catastrophes and advocates hasty actions.

As the debate unfolds, it has become increasingly clear that—contrary to conventional wisdom—*there does not exist today a general scientific consensus about the importance of greenhouse warming from rising levels of carbon dioxide.* On the contrary, most scientists now accept the fact that actual observations from earth satellites show no climate warming whatsoever. And to match this fact, the mathematical climate models are becoming more realistic and are forecasting temperature increases that are only 30 percent of what was considered the "best" value just four years ago.

We consider the Global Climate Treaty concluded in Rio de Janeiro at the 1992 "Earth Summit" to be unrealistic; its goal is stabilization of atmospheric greenhouse gases, which requires that fuel use be cut by 60 to 80 percent worldwide!

> Energy is essential for all economic growth, and fossil fuels provide today's principal global energy source.
>
> In a world in which poverty is the greatest social pollutant, any restriction on energy use that inhibits economic growth should be viewed with caution. For this reason, we consider "carbon taxes" and other drastic control policies—lacking credible support from the underlying science—to be ill-advised, premature, wrought with economic danger, and likely to be counterproductive.

These surveys and statements all confirm that many climate scientists believe that some global warming may occur in the future; but catastrophic predictions are unsupported by the scientific evidence, and even moderate warming forecasts are based entirely on yet-to-be validated climate models.

What do surveys mean in terms of greenhouse warming? Science is not democratic; truth is not arrived at by vote. *The surveys tell us that most scientists think there are still unanswered questions that need to be settled by additional research before drastic and far-reaching policies are undertaken. And there is time for this research* (see Box 9).

Climate Science Disputes

Temperature History

Temperatures over the last century are not in accord with what one would expect from climate model calculations based on the increase in GH gases since the Industrial Revolution. (Fig. 15)

The first problem certainly is the poor quality of the temperature data. Real geographic coverage has only been global since satellites went into operation in 1979. In addition, surface data are mostly land-based, nonuniformly distributed, and subject to local perturbations, principally the "urban heat island" effect (UHI) (Fig. 7). Ocean data are sparse, yet oceans account for 70 percent of the Earth's surface.

Box 9. Global Warming: Unfinished Business— the Need for Research

Climate science is not "settled"; it is both uncertain and incomplete. The available observations do not support the mathematical models that predict a substantial global warming and are the basis for a control policy on greenhouse (GH) gas emissions. We need a targeted program of climate research to resolve a wide range of unanswered questions.

A brief list of unsettled scientific issues:

1. The fate of anthropogenic CO_2 and its residence time in the atmosphere is uncertain: this includes its uptake into the ocean; the biological pump; the missing carbon sink. The future growth of atmospheric CO_2 depends on more precise estimates of residence time and the amounts of fossil fuels available for energy production. Some suggest a growth to eight times its preindustrial level, while others doubt whether CO_2 level will even double.

2. The temperature record of the last hundred years is of poor quality, with many discrepancies. Surface temperatures disagree with recent measurements from satellites and balloons. The "urban heat island" effect may skew the record. Ship observations are particularly questionable, with an unexplained abrupt cooling between 1900 and 1903.

3. General Circulation Models (GCMs) vary by some 300 percent in their temperature forecasts, require arbitrary adjustments, and cannot handle crucial mesoscale and microscale cloud processes. Their forecasts of substantial warming depend on a modeled positive feedback from atmospheric water vapor (WV) that may not exist.

4. GCMs cannot account for past observations: the unusual temperature rise from 1920 to 1940, the cooling to 1975, and the absence of warming in the satellite and radiosonde records since 1979. Various reasons for these discrepancies need to be explored before one places confidence in GCM forecasts of warming. Possible reasons are: reduced positive temperature feedback from water vapor; increase in cloudiness; the rise of anthropogenic aerosols; man-made changes of the land surface; increasing air traffic; solar variations influencing climate, and changes in oceanic circulation, particularly thermohaline-driven (i.e., by temperature and salinity gradients).

5. Data from tree rings, sediments, and ice cores all show prehistoric climate fluctuations on timescales as short as a decade; some data show changes larger than any present forecast for the future. These fluctuations are not explained by existing climate models, nor can GCMs account for El Niño events, the North Atlantic Oscillation, and other current rapid changes in climate.

6. Sea level (SL) rise is a major feared result of future warming. But increased evaporation from the ocean and more rapid accumulation of polar ice might lead to a lowering of sea level. This possibility is supported by an observed inverse correlation between SL rate of rise and tropical sea surface temperature (SST).

7. Severe storms and hurricanes (except in the eastern North Pacific), as well as precipitation, appear to have diminished in the past 50 years. A calculated global warming trend, primarily at high latitudes, would reduce the latitudinal temperature gradient and therefore the driving force for storms and severe weather.

8. Global agriculture may have benefited already from increased CO_2 and will continue to benefit from possible climate warming and increased precipitation, as increased nocturnal and winter warming leads to longer growing seasons. Increased CO_2 not only leads to more rapid growth of plants but also reduces their water requirements. Farmers can and will adjust to climate changes, whatever their cause.

9. The spread of disease vectors, like malaria-carrying mosquitoes, will be unimportant in comparison to the growth in human vectors. Medical science and better insect control technology will overcome disease encouraged by warmer climates.

10. Historical evidence supports the idea that warmer climate intervals are beneficial for human activities, food production, and health. Cold periods have had the opposite effect.

11. Mitigation techniques are available that can slow down the rise of atmospheric GH gases and possible climate change: energy conservation and increased efficiency often make economic sense; hydro and nuclear power are available now; solar energy is around the corner; tree planting and, especially, ocean fertilization may be low-cost methods of sequestering atmospheric CO_2.

12. Policy measures should be applied with great caution and only when justified by scientific data, lest they create more harm than good. Mandatory controls on energy use can create economic losses, harming especially poor people and poor nations.

Notwithstanding limited and questionable data, there is little doubt that global temperatures have increased in the past 100 years, well before GHG increased appreciably. Thermometers, tree ring data, and ice cores all show that an unusual warming took place starting in the last century and continuing up until about 1940. Many scientists would identify this warming as a natural climate variation, most likely a recovery from the deep cooling of the "Little Ice Age," perhaps due to a temporary acceleration of the thermohaline circulation and Gulf Stream drift into the North Atlantic.

The compilations of the global record differ depending on how land and ocean data are combined and what corrections have been made. For example, the compilation published by the IPCC differs from that of the Goddard Institute of Space Studies (GISS) (see Fig. 15). Both agree that temperatures rose between 1900 and 1940, but they disagree about the trends of the last 50 years. GISS shows a substantial decrease between 1940 and 1975, while the IPCC shows temperatures mostly rising since 1950. On the other hand, the compilation published by Hasselmann (1997) suggests a further increase beyond 1940. Tree-ring data also show a strong rise starting in 1880, but no temperature rise since about 1940 (Fig. 16).

Beginning about 1940, the climate cooled until about 1975 and then warmed abruptly. The overall change since 1940 may well be close to zero, depending on which climate record one accepts. There are various explanations for the cooling—from a natural climate fluctuation to increases in man-made aerosols that reflect some of the incident solar radiation, and to changes in the solar irradiance, the radiation flux emanating from the sun.

It is tempting to assign the sudden warming in the late 1970s to the enhanced greenhouse effect, but the time history of the temperature increase does not match the increase expected from the rise of atmospheric carbon dioxide. Nor is the climate history before 1975 consistent with enhanced greenhouse warming. The temperature history since 1979 is in dispute: satellite and radiosonde data show no warming, while surface data show increasing temperatures during the 1980s (Fig. 16).

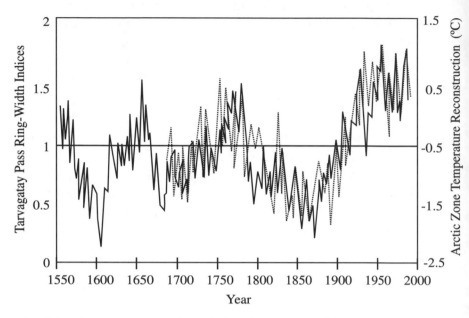

Fig. 16. The climate record as deduced from the width of tree rings. Compared are the ring-width chronology (solid line) and the reconstruction of Arctic annual temperature anomalies (dashed line) [Jacoby et al. 1996, reprinted with permission, © American Association for the Advancement of Science]. Note the sharp increase between 1880 and 1940.

There are different explanations for this discrepancy between satellite and surface data, but the key point is that the models calculate the temperature above the surface—which is what satellites and balloon-borne weather-radiosondes measure. This further sharpens the disagreement between models and satellite observations.

There is increasing evidence, however, that the surface data are contaminated by urban heating or other local influences. For example, a study (Karl and Jones 1989) on average temperatures in the continental United States, which carefully removes the UHI, does not support the common view of the 1980s as the hottest decade of this century. On the contrary, the temperature peak was reached around 1940. Similar results are reported from an analysis of European data (Balling 1997) (see Fig. 8).

An interesting analysis has been published by James Goodridge (1996), who plotted temperature trends over the last 90 years at over

100 California stations. Goodridge grouped the temperature trends according to population density. He found that temperature readings from counties of more than one million inhabitants show an "increase in temperature commonly attributed to greenhouse warming (as) 3.14°F per century" (see Fig. 7), while counties with less than 0.1 million people show essentially zero trend.[12] Significantly, the stations selected for a global compilation all show positive temperature trends (Fig. 17). This perhaps was due to a preference for stations located at or near airports, normally located at or near population centers. This selection introduces a bias towards stations affected by the UHI and may account for part of the discrepancy between "surface data" and the satellite data that truly measure an average global temperature unperturbed by urban heating (see also Fig. 18).

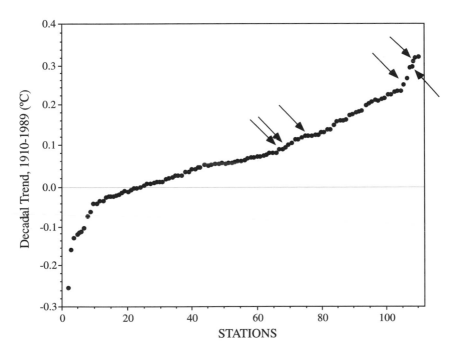

Fig. 17. Distribution of temperature trends for California weather stations. The arrows indicate the stations selected by GISS for a global temperature compilation [Christy and Goodridge 1995].

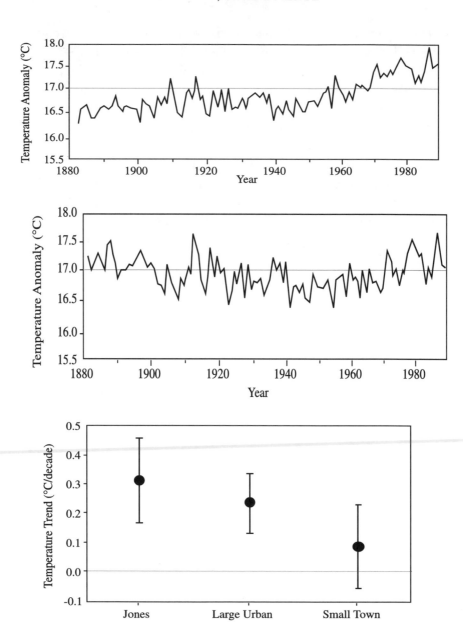

Fig. 18. A demonstration of the existence of the Urban Heat Island effect in Australia, comparing the record of (a) six capital cities with (b) 25 regional stations [Hughes 1992]. (c) Mean temperature trends for 1960–1990 for various South African data sets [Hughes and Balling 1997].

Why No Temperature Increase as Predicted by Models?

Once we demonstrate that the observations of the last 100 years are not consistent with model calculations, we need to find the reasons. Surely, the general inadequacies and shortcomings of the models themselves (see Box 10) illustrate the need for research to clarify the existing discrepancies between observations and theories.

Box 10. Shortcomings of GCMs
(All page references are to IPCC WG-I 1990.)

W. L. Gates (chief modeler, Livermore National Laboratory, Calif.), P.R. Rowntree and Q-C. Zheng:

"[Climate models] display a number of systematic errors in common....These model errors and sensitivities, and our current uncertainty over how best to represent the process involved, require a serious consideration of the extent to which we can have confidence in the performance of models on different scales." (p. 99)

"The existence of such common deficiencies, despite the considerable differences in the models' resolution, numerical treatments and physical parameterizations, implies that all models may be misrepresenting (or indeed omitting) some physical mechanisms." (p. 102)

G. McBean (head of the Canadian Weather Service) and J.M. McCarthy (Harvard University):

"Present [model] shortcomings include: Significant uncertainty, by a range of three, regarding the sensitivity of the global average temperature and mean sea-level to the increase in greenhouse gases. Even larger uncertainties regarding regional climatic impacts, such that current climate change predictions have little meaning for any particular location...." (p. 317)

"If the atmosphere and upper-ocean alone were responding to the increase in greenhouse heating and the cloud-radiation feedback operated according to current knowledge, then the surface of the earth would already be 1 to 2 degrees C warmer than the temperatures of the nineteenth century." (p. 321)

We will discuss four different classes of effects that could, in principle, explain why the increased radiative forcing by CO_2 has not produced a substantial temperature increase, and in particular, why the satellite data since 1979 show no upward trend:

— An (internal) negative feedback counteracting the warming from increased GHG, involving the atmosphere itself or the coupled atmosphere-ocean system.

— Increased negative radiative forcing produced by (mainly sulfate) aerosols caused by human activities related to energy generation and CO_2 emission.

— Radiative forcing produced by human activities not directly related to energy production.

— Solar effects or natural climate fluctuations that fortuitously offset a warming by the enhanced greenhouse effect.

A principal task for climate research is to gain information about the relative importance of these different classes of mechanisms so that climate models can be improved and produce a better simulation of the real atmosphere. The most promising technique involves looking for "fingerprints" in the climate record, i.e., the geographic, vertical, and temporal variations of temperature or other important climate parameters. These can then be compared to predicted changes for each of the mechanisms.

Cloud feedback: The most obvious negative feedback effect would come from increased cloudiness. It has been assumed for some time, and by various authors, that a warming of the ocean surface would lead to increasing evaporation, increased cloudiness, increased albedo—and thus a negative feedback that cancels much or most of the warming (see Box 11).

Reduced positive (or even negative) WV feedback: Water vapor in its various forms accounts for more than 95 percent of the natural greenhouse effect. Indeed, the large temperature increases predicted by the standard GCMs come about because they incorporate an effectively constant relative humidity, i.e., as atmospheric temperatures rise the WV content also increases—and so does its radiative forcing. Without the positive feedback built into most climate models, the calculated temperature increase due to a doubling of CO_2 would only be 1°C.

Box 11. Clouds Are Crucial to Climate Predictions

Bruce Callander, head of the IPCC Science Panel, was quoted in the *London Financial Times* (18 March 1994, p.14):

Cloud behaviour is the "single biggest uncertainty." Researchers cannot be certain whether (clouds) speed warming or slow it...in 10 years we may say (scientists' investigation) has been an interesting exercise which came to nothing, or we may say that we were recognizing something important happening in the atmosphere.

Noted cloud physicist Peter Hobbs (1993) of the University of Washington says:

"...in the absence of an understanding of the physical processes that control climate and their adequate representation, models are unlikely to provide reliable predictions of climate change."

University of Arizona atmospheric physicist Dale Ward finds that climate models overestimate the amount of solar energy reaching the earth's surface by up to 50 percent, mainly because of their inadequate treatment of the optical properties of clouds.

Meteorologist Richard Somerville of Scripps Institution of Oceanography, noting the discrepancy between data and theory, asks:

"Have the clouds meanwhile changed so as to amplify or reduce the resulting climate change? We do not know. In fact, we lack a basic understanding of why the global cloud amount is about 60 percent, why the planetary albedo is about 30 percent, how these and other fundamental quantities may have changed along with climate over geological time, and how they may change in the future."

The models, so far, cannot answer any of these questions.

Principally Ellsaesser (1984) and Lindzen (1990) have pointed out that reducing water vapor content in the colder upper troposphere could lead to a negative feedback that reduces overall GH warming. Several mechanisms have been identified that could lead to such drying as a result of enhanced convective activity at the surface (Sun and Lindzen 1993). Recently, Broecker (1996) has taken up this idea and tried to explain rapid climate changes in terms of changes in atmospheric water vapor. Clube et al. (1997) have also pointed to water vapor changes to produce large climate effects.

An independent physical approach leading to a similar conclusion, has been followed by Hugh Ellsaesser (1984, 1989, 1990). He argues that increased radiative forcing, by heating the tropical region, enhances convective activity and strengthens the Hadley circulation. In turn, the increased subsidence at extra-tropical latitudes causes a drying of the upper troposphere there, leading quite naturally to a negative feedback (see Box 12).

Observations have been difficult, particularly on a global scale. Spencer and Braswell (1997), however, have demonstrated that satellite microwave and IR data can be used to obtain global WV distributions in the upper troposphere.

Box 12. IPCC's Evolving Understanding of Water Vapour Feedback (1990-1996)

"The best understood feedback mechanism is water vapour feedback, and this is *intuitively easy to understand.*" (IPCC WG-I 1990)

"There is no compelling evidence that water vapour feedback is anything other than positive... *although there may be difficulties with upper tropospheric water vapour.*" (IPCC WG-I 1992)

"Feedback from the redistribution of water vapour *remains a substantial uncertainty in climate models.*... Much of the current debate has been addressing feedback from the tropical upper troposphere." (IPCC WG-I 1996)

Aerosol cooling: The large discrepancy between the predictions of conventional GCMs and the climate record of this century has focused attention on aerosols. Their negative radiative forcing (by scattering incoming solar radiation and thereby increasing albedo) would mitigate greenhouse warming and may even produce a climate cooling.

But it is becoming clear that the validation of aerosol-enhanced GCMs may be rather difficult. It is not enough simply to reduce the positive radiative forcing of greenhouse gases, of about 2.5 Watts/m^2, by the cooling effect of sulfate aerosols. There are major problems here, and a number of minor ones, all leading to uncertainties of forcing of more than a factor of ten!

For example, Tett et al. (1996) have modeled the direct albedo effects of sulfate aerosols and find that their influence on calculated temperatures over the period 1961 to 1995 is hardly detectable. Schwartz and Andreae (1996) stress the uncertainties due to aerosols. When they consider the aerosols' indirect effects (i.e., through nucleating clouds and thereby increasing albedo further), the global mean radiative forcing can vary between about -0.6 up to +4.0 Watts/m^2 (Fig. 5). But because of the peculiar geographic distribution of aerosols, whether sulfates, soot, or mineral dust, "global mean" is not really the significant parameter here and can in fact be quite misleading.[13]

Hansen et al. (1997) have published a comprehensive analysis of the radiative forcing effects of aerosols and of other factors, such as changes in solar irradiance, ozone, clouds, and well-mixed greenhouse gases; they have investigated the influence of arbitrary heights, latitudes and longitudes, seasons, and time of day to see how this affects the climate response. Contrary to the major thrust of the Second Assessment Report of the IPCC (1996), they conclude that anthropogenic aerosols are of lesser importance, but that the ozone depletion of the last 20 years can cancel up to 50 percent of the GH gas climate effects.[14]

Other human influences on the atmosphere and climate: The major human influence certainly is energy generation, the burning of fossil fuels, which produces carbon dioxide and various pollutants, some of which can lead to aerosols. In addition, there are the following human-related processes affecting climate:

a. Methane (CH_4) production, principally by cattle raising, rice growing, land fills, oil and gas operations

b. Nitrous oxide (N_2O) production from fertilizer in agricultural operations

c. Other GH gas releases into the atmosphere, such as CFCs

d. Heating in populated regions, mainly by the release of energy

e. Changes in surface albedo by land clearing, agriculture, roads, etc.; creation of mineral dust

f. The increased evaporation of WV by agriculture, irrigation, man-made reservoirs, and air-conditioning

g. Large-scale burning, creating smoke, soot, and various polluting gases

h. Diversion of fresh water from the Mediterranean (with consequences to the Atlantic circulation)

i. Air pollution leading to smog and low level ozone

j. Increased tropospheric ozone (may not be man-induced)

k. Decreasing stratospheric temperatures (may not be man-induced)

l. Increasing stratospheric sulfates (may not be man-induced)

m. Increasing stratospheric water vapor

n. Decreasing stratospheric ozone

o. Rapidly increasing air traffic

Some of these changes have causes that are easy to recognize. For example, as human population grows in developing nations, agricultural activities and deforestation change surface albedo. The increase in atmospheric methane is also related to population growth and agricultural activities (rice growing and cattle raising).

Some of the observed stratospheric changes are undoubtedly due to increased air traffic penetrating into the lower stratosphere, injecting sulfur dioxide, nitrogen oxides, as well as water vapor, leading to

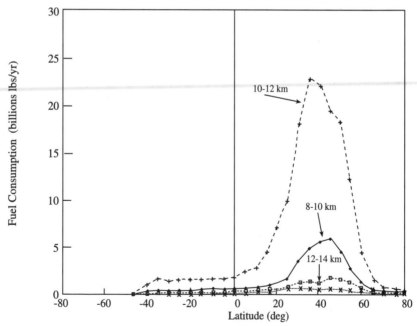

Fig. 19. Fuel consumption for 1990 air traffic as a function of altitude and latitude (summed over longitude) derived from NASA data [Singer 1997b].

contrails and invisible cirrus with strong IR effects (Fig. 19). Results from air traffic could indeed produce detectable climate effects on the Earth's surface: a heating trend at northern midlatitudes, as observed in the satellite temperature record (Fig. 9).

Many of these effects have so far been ignored in IPCC reports. If future research bears out their importance, it would underline further the inadequacy of current climate models, which do not include many potential human influences on climate.[15]

Solar variability: It has long been suspected that changes in the energy output of the Sun would change the climate. Various authors have linked observed climate parameters to the solar cycle, and temperature changes to the variation of solar-cycle length (Fig. 3). Convincing evidence shows that atmospheric circulation changes are linked to the solar cycle (Labitzke and van Loon 1994). And it is well known that total column ozone shows a 2 to 3 percent solar-cycle variation at midlatitudes, and an even larger variation at higher latitudes.

The scientific community has been slow to accept the existence of a solar influence on climate in spite of excellent statistical correlation (Friis-Christensen and Lassen 1991; Baliunas and Jastrow 1990). The conventional position is that the variation of the solar constant is too small to have an effect, only 0.1 percent during the solar cycle. But the variability is much greater in the ultraviolet, as can be deduced from both the observed ozone variation and from direct observations above the appreciable atmosphere (Lean 1991). While the energy content of the solar UV is small, its effect on the ozone layer could influence the atmospheric circulation (Haigh 1996). Solar variations also produce indirect climate changes through solar corpuscular radiation (solar "wind") sweeping past the Earth and the solar modulation of the flux of cosmic rays (Svensmark and Friis-Christensen 1997) (see Fig. 10).

What Can We Learn from the Past Climate Record?

Independent research must involve a study of the paleoclimate record, extending into the past the record from surface thermometers. It must involve analysis of the width of tree rings, isotopic composition of ice cores, study of ocean sediments, and other means of gaining information. An impor-

tant finding is that sudden changes (on time scales as short as decades) have occurred during the present interglacial period (Holocene) and during past glacial maxima (Fig. 2). These changes are often correlated with changes in greenhouse gases, although these may not be the cause of the temperature changes. So far, GCMs cannot account for such natural changes.

Both paleoclimatological and historical data confirm a warm period, the Medieval Climate Optimum (900–1300 AD), and a cold period, the Little Ice Age (1450–1850 AD), with temperature departures about ± 1 deg C (1.8 deg F) from the mean of the Holocene. Since we have no explanation for these oscillations—and even larger ones in the past (Keigwin 1996)—we must accept the possibility that they will continue. Even IPCC WGI (1990, p. 203) admitted: "...some of the global warming since 1850 could be a recovery from the Little Ice Age rather than a direct result of human activities." If the climate is indeed moving out of the Little Ice Age and possibly into another climatic optimum, like that which preceded it, the roughly 0.5 deg C warming of the past 130 years of observational record is evidence of neither greenhouse warming nor anthropogenic effect. We could even look forward to an additional warming of about 1 deg C (1.8 deg F) over the next few centuries regardless of what man does to prevent change.

Severe Storms Not Increasing

Studying past weather and storm patterns can give us a standard for evaluating the effects of a possible future warming. Both theory and observations agree that severe storms, both extratropical and tropical, have not increased in the past 50 years. In fact, North Atlantic hurricanes have noticeably declined in frequency and in intensity (Landsea et al. 1996) (Fig. 20).

Precipitation Not Increasing

Climate models all predict an increase in precipitation as a result of GH warming, yet research shows that global land precipitation ceased to increase circa 1950 and has decreased since, except at high northern latitudes (IPCC 1996). So far, we have no satisfactory explanation for this fact.

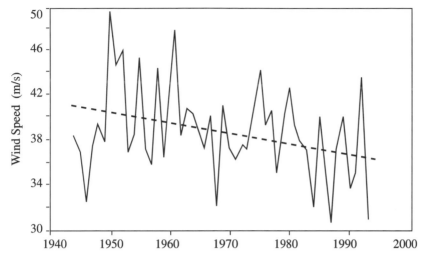

Fig. 20. Time-series of mean annual maximum sustained wind speed attained in Atlantic hurricanes [Landsea et al. 1996]. Linear trend shown as dashed line [IPCC 1996].

Increase in Future Sea-Level Rise Doubtful

Climate models predict that warming will increase evaporation, and therefore precipitation, although the geographic distribution cannot be established with any certainty. If precipitation increases over the polar regions, then ice will accumulate on Antarctica and Greenland and lead to a lowering of sea level. On the other hand, a global warming will lead to some increase in sea level because of thermal expansion of ocean water and melting of glaciers; the current observations do not give us any sure guidance on this important question. Sea level has been increasing during the past centuries for reasons that may have nothing to do with climate change (Fig. 21). The detailed changes of sea level measured during this century, however, suggest an anti-correlation with global temperature (Singer 1997a) (Fig. 11). By this measure, *global warming should therefore lead to a slow-down rather than an acceleration in sea-level rise.*

Agriculture and Other Human Activities Will Benefit

The greatest impact of climate change, historically, has been on agriculture, with warm periods producing larger harvests and cold periods causing famines. A sustained warming, particularly with a reduced diurnal and seasonal temperature range, with warmer nights and milder winters, should benefit agriculture by extending the growing season—already suggested by observations of the changing seasonal cycle of CO_2 (C. D. Keeling et al. 1996). This will be aided by the fertilizing effect of carbon dioxide and the reduced need for water by plants.

Other impacts of a sustained warming should also be beneficial; for example, historic records show positive impacts of warm periods on human health (Moore 1995). The current concerns about the spread of tropical diseases are certainly overblown when we consider that the most important vector in the spread of diseases is the human vector, aided by the growth and rapidity of global transportation. It would not be an exaggeration to state, as does Thomas Gale Moore (1995) and others, that global warming is good for you.

In summary, temperature observations of the past century, and especially satellite and radiosonde data since 1979, do not suggest a warm-

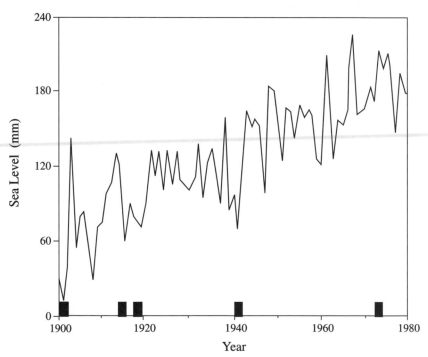

Fig. 21. Sea-level trends for 84 stations with more than 37 years of data [Trupin and Wahr 1990], corrected for post-glacial rebound. Occurrences of major El Niño events are indicated on the time axis; they generally correlate with dips in sea level. Compare with Fig. 11.

ing caused by human activities; thus the climate models that predict a major warming in the 21st century are not yet validated and should not be used for policy purposes. At most, we believe there will be a modest warming in the next century, generally beneficial for agriculture and human welfare. Available evidence suggests that none of the extreme fears about severe weather events, sea level rise, and spread of diseases, are warranted.

Mitigation of GHG Increase and of Climate Changes

A variety of schemes have been offered for counteracting any GH warming of the planet; some rely on reducing the amount of solar energy reaching the surface. Others try to reduce the atmospheric levels of CO_2, the major anthropogenic GHG, by speeding up its absorption. The most controversial policies involve reducing the emissions of CO_2 by controlling the burning of fossil fuels without any specific plans for substituting other energy sources (see Appendix).

A very different strategy, favored by many economists, is to strengthen the economies, particularly in developing nations, to enable populations to adjust to any likely climate change. Adaptation to climate change, seasonal and interannual, has been the rule throughout human history; populations have experienced and adapted even to major long-term changes in climate when they migrated over large distances.

Increasing Earth's Albedo?

Fanciful schemes have been suggested for reducing the amounts of solar energy reaching the surface. Some involve shooting aerosols into the stratosphere, producing a cooling analogous to that following volcanic eruptions. Putting "venetian blinds" into orbit seems bizarre and not technically feasible in the foreseeable feature (National Academy of Sciences 1991). But there are no scientific obstacles to these ideas, only lack of resources. Perhaps in the future, solar energy collectors, made from material mined on the Moon, could fulfill a dual function of generating electric power and at the same time decreasing (or increasing, if need be) the amounts of solar energy reaching the Earth.

Sequestering Atmospheric Carbon Dioxide

Planting trees is a known method of reducing the CO_2 content of the atmosphere, but the resources required in terms of land areas and effort are immense. Since the trees eventually die, the CO_2 is not removed permanently, but its growth rate is reduced and so is any putative climate impact. Tree planting spreads out the CO_2 peak over centuries, allowing the natural but slow absorption of excess CO_2 by the oceans to limit the CO_2 concentration. Preferential incorporation of more wood into new permanent structures would accelerate biospheric CO_2 storage.

If recent experiments are any indication, fertilizing the oceans with iron, an essential micro-nutrient, will permit greatly increased growth of phytoplankton, which will eventually transform more atmospheric CO_2 into carbonates that sink to the ocean bottom—accelerating the "biological pump." As discussed in the Appendix, ocean fertilization may constitute the least-cost approach to GHG mitigation, one that is vastly less expensive than controlling emissions.

Reducing GHG Emissions

There is no dispute that levels of greenhouse gases (carbon dioxide, methane (CH_4), nitrous oxide (N_2O), CFCs, etc.) have increased as a result of human activities. Carbon dioxide (CO_2) has been increasing at about 0.5 percent per year, mostly as a result of fossil-fuel combustion, related directly to energy generation. Its level, 360 ppmv (part per million by volume), has increased by nearly 30 percent over the preindustrial level of 280 ppmv. The sources of methane, whose concentration has doubled in the last 100 years, are more varied: in addition to natural sources from swamps and wetlands, human sources (growing as population and economic activities increase) include fossil-fuel operations and landfills, but also cattle raising and rice growing.

The most important atmospheric greenhouse gas is water vapor; but almost all policy proposals for emission control are focused on anthropogenic carbon dioxide. Our industrial society now pumps about 6 gigatons (billion tons) per year of carbon into the atmosphere. (1 Gt

= 10^9 tons = 10^{12} kg = 10^{15} gm.) What happens to it? The fate of the CO_2 in the atmosphere needs to be understood before far-reaching control measures are instituted.

We know from the observed rate of increase that over 40 percent of anthropogenic CO_2 remains in the atmosphere. The remainder, we think, is absorbed by the surface layers of the ocean, eventually to be transported into the deep ocean over a period of centuries. Measurements indicate, however, that much less than 60 percent is absorbed, creating the problem of the missing "carbon sink." The latest research (R.F. Keeling et al. 1996) provides a likely answer, but still may not be the final word. The missing sink is identified with uptake of CO_2 by forests and soils, mainly in the northern hemisphere. The increasing amplitude of the seasonal cycle of CO_2 strongly suggests that the biospheric mass is increasing.

There is growing evidence for the existence of a CO_2-fertilization effect, increasing the amount of biomass as CO_2 levels rise. Such a development would increase the fraction going into biomass and decrease the fraction of emitted CO_2 remaining in the atmosphere below the current 40 percent, thus slowing the rate of growth of atmospheric CO_2. This "negative feedback" effect has not been considered explicitly in the CO_2 projections of the IPCC report.

How Will GHG Increase?

There are three areas of uncertainties in these estimates, particularly as concerns CO_2, which need more detailed analysis: 1. residence time of CO_2 in the atmosphere; 2. emission rates over time; and 3. the total amount of fossil fuels available and the amount actually used (see Box 13).

If a GHG has an atmospheric lifetime of one or more centuries, then any emission whatsoever will increase its concentration.[16] The lifetime of carbon dioxide is estimated at 50 to 200 years; but its cumulative emission over the next few decades will determine its concentration. We agree with the IPCC assessments that a cumulative emission between the period 1991 to 2100 of 630 gigatons of carbon (GtC), 1,030 GtC, and 1,410 GtC, would lead to CO_2 levels of 450, 650, and 1,000 ppmv, respectively.

Box 13. Review of IPCC Emission Scenarios (B.J. Mason)

From a review by B.J. Mason (1995), former director of the British Meteorological Office, of the IPCC report, *Climate Change 1994: Radiative Forcing of Climate and an Evaluation of the IPCC 1992 Emission Scenarios.* Cambridge University Press, 1995.

"Any emission scenario covering the next 100 years can, at best, be little more than an intelligent guess based on future projections of global population, GDP, energy demand, new technological developments, governmental action and public response—all of which are very difficult to predict and involve great uncertainty. For instance, population projections for the year 2100 in the various emission scenarios range from 6.4 to 16 billion! It is therefore hardly surprising that the scenarios, starting from different initial assumptions, diverge sharply with time to predict CO_2 emission rates for Y2100 that range 50-fold, from 1.2 to 60 GtC, with 7-fold differences in cumulative totals. Faced with this problem, the authors adopt a policy of nondiscrimination and merely advise the use of many different scenarios to assess climatic, economic and social impacts—but to what purpose? Given the poor track records of economic and energy forecasts, even on the near term, there seems little point in modifying emission scenarios at frequent intervals, and even less in offering an even wider range of projections with even greater levels of uncertainty. This will encourage an even greater output of speculative papers that cannot be profitably compared with reality or even with each other."

To put these numbers into perspective: Stabilization of CO_2 at the current level (360 ppmv) requires that emissions, *worldwide*, be reduced by 60 to 80 percent of 1990 levels, according to IPCC WG-I (1990 and 1992). These numbers are avoided by politicians. They prefer to discuss stabilization at 550 ppmv, which requires a reduction of about 50 percent.

Perhaps the greatest uncertainty relates to the scenarios involving the use of fossil fuels over the next century, especially in rapidly growing emerging nations like China, India, etc. (see Fig. 22). The uncertainties of fuel use are so large that they overwhelm that of CO_2 lifetime, if we ask the question about a CO_2 level in, say, the year 2100. On the other hand, if we ask the question differently, for example, will CO_2 concentrations double (at any time), then the lifetime and physical processes become more relevant (Gerholm 1992; Linden 1998).

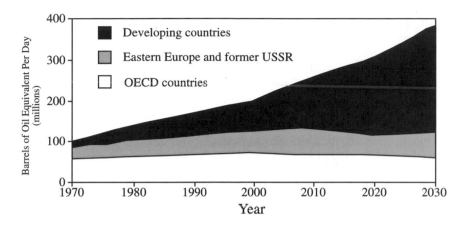

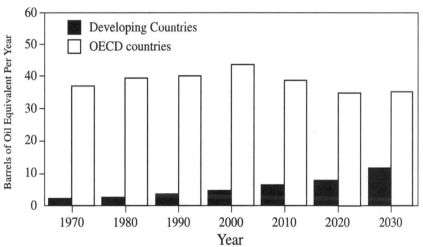

Fig. 22. (top) Energy consumption by country group, roughly equivalent to CO_2 emission. By the year 2000, the output from non-OECD countries will predominate. Marland et al. [1994] note that emissions from OECD have nearly leveled off, while emissions from China have been doubling every 20 years since 1960. (bottom) Per-capita consumption by OECD and developing countries. After the year 2000, OECD consumption diminishes. The projections assume an "energy-efficient" scenario, i.e., with growth in total consumption of 1 or 2 percentage points below the trend rate [adapted from 1992 World Bank data; see also Lichtblau 1998].

Economic Considerations: Costs and Benefits

The discussion of emission scenarios is outside the realm of climate science and should be considered by specialists. It seems clear to us now that if emission reduction is postponed to a later date, economies will benefit for several reasons:

1. Existing capital stock will be allowed to wear out naturally before being replaced.

2. Because of the time-value of money, deferred investments in capital equipment are preferable.

3. New technologies will be developed, whether higher-efficiency or renewable-energy, that will permit a lower-cost solution.

4. We may learn more about climate science and find that a large emission reduction is unnecessary.

Minimizing the cost of emission controls

There are two basic methods for reducing the overall costs of reducing CO_2 emissions without substantially affecting the eventual concentration and putative climate effects. One method is to share the burden internationally, reducing emissions at the least cost by emission trading. This would mean establishing a market for emission rights. The problem with this approach is obvious, namely how to assign initial emission rights. If national quotas were to be established, would they be based on population? If so, should they be based on present population or some future figure? A parallel problem exists for per-capita energy consumption. Since these decisions affect the flow of multibillions annually, they are certain to be highly controversial.

Another approach is to vary the timing of emission controls. By postponing emission cuts, it is possible to take advantage of improved technology—new capital equipment of higher efficiency or the development of energy sources that do not emit CO_2. Because of the long atmospheric lifetime of CO_2, it is the cumulative amount emitted that determines atmospheric concentration—to a first approximation. Once the cumulative amount has been agreed on, the time pattern of emission can be optimized for least cost. This possibility has been explored in a study by Wigley, Richels and Edmonds (1996), which develops a set of illustrative pathways for stabilizing atmospheric CO_2 concentration at 350-750 ppmv over the next few hundred years (see Fig. 23).

Uncertainties in economic estimates

An up-to-date summary of the monetized damage from a doubling of CO_2 to the U.S. economy is given in Table 6.4 of WG III (IPCC WG III 1996). The discrepancies among individual authors are quite large, however. For example, for agriculture, Nordhaus and Titus give an annual damage of $1.1 and 1.2 billion (for a temperature increase of 3°C and 4°C, respectively), while Cline and Tol calculate figures of $17.5 and 10 billion (for a temperature rise of 2.5°C), respectively. The discussion has concentrated almost entirely on defining and calculating damages produced by higher CO_2 levels and by warming. Little attention has been paid to the benefits (see Box 4).

The uncertainties in calculating control costs are well illustrated in the same report (IPCC WG III 1996, fig. 9.1), where some authors calculate a GNP loss for the U.S. of 2 percent for a 10 percent CO_2 emission reduction, while others calculate only a 3 percent loss for a 90 percent reduction.

A least-cost approach to CO_2 control

The three Working Groups of the IPCC, in their separate volumes, consider, respectively, the science, the impacts and mitigation, and the economic and social dimensions. A great deal of their discussion is devoted to the control of emissions of greenhouse gases. WG II and WG III also consider in some detail the feasibility and costs involved in afforestation as a means of sequestering atmospheric CO_2 and thereby reducing its concentration.

Curiously, none of the three volumes consider ocean fertilization as an alternative or as a supplement to emission control or afforestation. IPCC WGI (1996) is rather pessimistic about the feasibility of ocean fertilization, but their assessment may not have been aware of the successful test (Cooper et al. 1996). Yet from the preliminary analysis shown in the Appendix, ocean fertilization may constitute the least-cost approach to CO_2 control. The cost becomes even lower, and may in fact be negative, if one considers the value of fisheries resources that might be created by ocean farming.

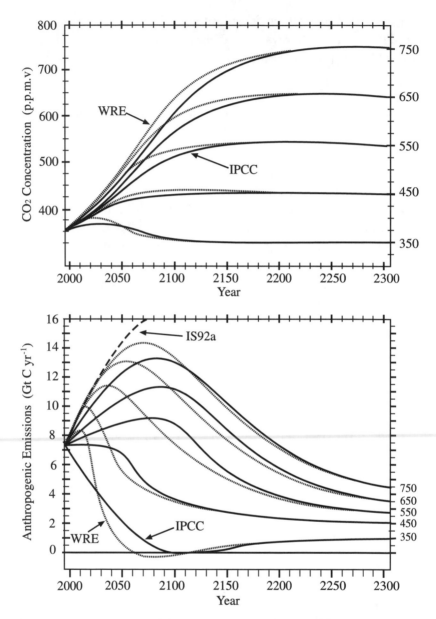

Fig. 23. A comparison of IPCC [1994] scenarios for anthropogenic CO_2 emissions with economically optimized emission scenarios [adapted from Wigley, Richels and Edmonds 1996]. The WRE scenarios permit higher emissions in the short run but require lower emissions in the medium term, around the year 2100, in order to achieve similar final CO_2 concentrations. (IS 92a refers to the IPCC [1990] "business-as-usual" emission scenario—actually unrealistically high.)

What should be the goal for CO_2 stabilization?

This brings us to the final point. Article 2 of the FCCC calls for stabilization of greenhouse gases at a level that will "prevent dangerous anthropogenic interference with the climate system." The position of such a level has not yet been established scientifically. Presumably, "dangerous" refers to a climate change that is unanticipated by current models, which are essentially "linear" (in the sense that more CO_2 translates into a larger temperature rise).

There is no reason to believe that the present CO_2 level of 360 ppmv is less dangerous than a level of 500 ppmv or 1000 ppmv. After all, about 500 million years ago, the Earth experienced CO_2 levels about 20 times the preindustrial value of 280 ppmv (Berner 1997) (see Fig. 1).[17] For that matter, we do not even know whether the present level is more dangerous than the preindustrial level.[18] We do know that it is not feasible to achieve a preindustrial level unless we go to *negative* emissions. Even maintaining the present level requires that *worldwide* emissions be cut by 60 to 80 percent; this reduction would have drastic economic effects, if indeed it could be accomplished. *Once we adopt the argument, however, that only politically feasible solutions are possible, we might as well adopt a conservative and prudent approach based on economics and cost-benefit analysis, rather than on exaggerated fears.*

One often hears the hypothetical question, "How do we know that there will not be a climate surprise?" There could be a runaway climate warming, or other catastrophe, once a certain "dangerous" GHG level is reached, one fears. But we know that CO_2 levels have been many times higher in the past than today's value without causing irreversible climate catastrophes. On the other hand, major and rapid temperature changes have occurred not only during glacial periods of the last 2 million years, but even during the Holocene of the past 11,000 years.

Finally, we must address the "precautionary" argument which says that even if we don't know that there is a problem, we still must deal with it. This approach is simply not realistic or logical. People will not buy insurance and pay high premiums for negligible dangers. No one seems willing to spend major sums to protect the Earth against the truly catastrophic impact of an asteroid or comet—which is sure to happen eventually, if past experience is any indication.

The Kyoto Protocol Is Ineffective

The Kyoto Accord, concluded in December 1997, would, if ratified, oblige Annex-I nations to reduce greenhouse gas emissions by an average of 5.2% (with respect to the base year 1990) by the period 2008 to 2012. The Accord was finally signed by the United States in November 1998, over the vociferous opposition of many members of the US Senate. (In July 1997, the Senate had passed a Resolution against a Kyoto-like agreement by a vote of 95:0 — principally because it would not include some 130 developing nations, including China, India, and Brazil.)

Even if all nations were required to follow the Kyoto Accord, it would merely slow down somewhat the ongoing rise in atmospheric GHG concentration. As calculated by the IPCC, a worldwide emission cut of 60-80% is required to stabilize CO_2 at present levels. China, now the world's third-largest CO_2 emitter, has been doubling its emissions every 25 years for the past 50 years, and is likely to continue for some time to come. Limiting emission cuts to only industrialized nations would create great inequities and result in economic disaster for most nations, including those that rely on trade with industrialized countries.

In any case, the climate effect of a Kyoto Accord is minute. If one were to accept IPCC figures, the additional warming, predicted for 2050, of 1.39°C would be reduced to 1.33 °C , a reduction of only 0.06°C (Parry et al. 1998). A 20% cut in emissions would yield a temperature of 1.22°C; a cut of 30% (six times the Kyoto target) only 1.19°C.

3

Epilogue

What to Do about Greenhouse Effects

Adapt to Climate Change and Apply a "No-regrets" Energy Policy

In 1989, then-Senator Albert Gore told *Time* magazine for its "Planet of the Year" issue "...that we face an ecological crisis without any precedent in historical times is no longer a matter of any dispute worthy of recognition," and "those who, for the purpose of maintaining balance in the debate, take the contrarian view that there is significant uncertainty..." are like people who permitted (by ignoring the signs) the murder of millions by Nazi Germany. More recently, Vice President Al Gore (1995) has accused scientific skeptics of treating global warming as the "empirical equivalent of the Easter Bunny."

The fact that human activities are changing the chemical composition of the atmosphere has evidently become a cause of concern for much of the public. The first question to ask is: What are the physical consequences of this "grand geophysical experiment"—as Professor Roger Revelle, the "father" of greenhouse warming (and Al Gore's mentor), referred to it. How significant are its effects on global climate, ecological values, and human welfare? The second question is: What, if anything, can and should be done about it?

With regard to the first question, there is the extreme, alarmist position that visualizes concentrations of greenhouse gases, principally carbon dioxide (CO_2), increasing to levels that have not been experienced on the earth since the age of the dinosaurs, about five times the present level. The consequences, based on unverified theoretical climate models and fanciful speculations, are pictured to be catastrophic to humans, ecosystems, and even to the climate itself. At the other extreme are those who would deny that human activities can produce changes in the global environment.

A rather different approach is to pay heed to scientific evidence rather than opinions. It is certainly significant (1) that climate data show no current warming (Spencer and Christy 1992)—in sharp contrast to the results from computer models—in spite of a rise in greenhouse (GH) gases equivalent to a 50 percent increase in CO_2 in the past century; (2) that in the last several thousand years there have been natural climate variations larger and more rapid than those predicted by computer models (Keigwin 1996); (3) that any future warming may result in a slowdown, not an acceleration, of the ongoing rise in sea level (Singer 1997a); and (4) that rising atmospheric CO_2 concentrations are a boon to agriculture (Idso 1989, 1995).

It should be recognized, perhaps, that there is a consistent bias in the public's understanding of GH warming and most other environmental problems. Nearly all research is funded by government; the rationale for expending taxpayers' money is to prevent hazards to the general population. Funding agencies therefore do not look kindly on research proposals that could demonstrate environmental "hazards" to be not serious or even nonexistent. But this means that essentially no research funds are provided to document and quantify some of the undeniable benefits from man's impact on the environment (see Box 4).

Panicky versus Prudent Measures: The Case for Adaptation

With regard to the second question, activists and most bureaucrats want to stabilize the atmospheric concentration of CO_2—even if this requires a worldwide reduction in present emission levels and energy use by 60 to 80 percent! They visualize an elaborate scheme of setting national energy quotas, essentially a rationing system. Superimposed would be a legal "black market" that would permit the buying and selling of emission rights between nations; it is equivalent to imposing onerous taxes on energy users, with revenues flowing to other nations' treasuries. Even without national and international taxes, such an emission control scheme would depress GDP by several percent, impose costs on the world economy in the multitrillion dollar range, and cause hardships for low-income groups and the poorest nations (Manne and Richels 1991).

Since we have no scientific guidance on what constitutes a "dangerous" level of greenhouse gases (referred to in Article 2 of the Global Climate Treaty), a more reasonable position is to institute prudent, "no-regrets" measures that make sense even without the threat of global warming disasters. Such measures, many of which have already been put in place, include sensible energy conservation and efficiency. (Electric power plants are continuing to show higher conversion efficiencies; in future, hybrid electric cars, using regenerative braking, might easily achieve twice the fuel economy of the present automobile fleet.)

A complementary approach, about which there has been amazing silence, is to encourage the use of nuclear and other alternative energy sources, and mitigate any negative effects of climate change in the same way one tries to mitigate the effects of the much larger natural climate variations.

The most reasonable policy, then, is to adapt to climate change, as human activities normally adapt to seasonal and year-to-year variations. One can then use the funds saved to strengthen the resilience of national economies, particularly in developing countries, against naturally occurring extreme climate events that cause damage (Goklany 1992, 1995).

Sequestering Atmospheric CO_2

Many energy specialists consider the period of fossil-fuel use—and elevated CO_2 levels—to be limited to one or two centuries, a mere blip on the time scale of human existence. Specialists express doubt that atmospheric CO_2 would even double (Linden 1998; Gerholm 1992). Still, given the many uncertainties about how climate change comes about, it makes sense to speed up the natural absorption of excess atmospheric CO_2 into the biosphere and ocean.

With the residence time of atmospheric CO_2 variously estimated as between 50 and 200 years (IPCC WG-I 1996), its current excess over its preindustrial value will eventually be absorbed by biota on land and in the ocean anyway. But even if a future warming is negligibly small and on the whole beneficial, there may still be political pressure to control the level of atmospheric carbon dioxide. It is hard to imagine broad political support for an emission control plan, given the disastrous economic consequences it entails. Fully realizing this, politicians have instead talked about more modest reductions of between 5 percent and 20 percent from current rates—with more to come later. But even if these reduced rates were to be achieved on a worldwide basis, they would do no more than to slow down somewhat—and at great cost—the current upward trend of atmospheric CO_2. *Stabilizing emissions does not stabilize concentration if the atmospheric residence time is so long that CO_2 accumulates.*

An alternative approach to emission control is to sequester the CO_2 from the atmosphere—or at least demonstrate that sequestering is technically and economically feasible. The conventional approach to CO_2 sequestration has called for tree planting on a massive scale, thereby tying up CO_2 for decades, to be released when the wood decays (IPCC WG-II 1996). But tree planting can be costly and impractical; it requires huge areas and great expenditures of funds to make an appreciable impact. Order-of-magnitude figures for sequestration by trees are 0.8–1.6 tons of carbon per hectare per year (Nordhaus 1991a, 1991b); thus, to absorb current production of about 6 gigatons CO_2 requires about 50 million square kilometers (ca. 4,500 x 4,500 miles!). One relatively low-cost policy, however, would be to use as much lumber as possible in all permanent structures and reseed existing forests.

Ocean Fertilization: A Scheme to Draw Down Atmospheric Carbon Dioxide

A technique analogous to afforestation, but economically more attractive, is to speed up the natural absorption of CO_2 into the ocean. Currently, much of the world's oceans is a biological desert. Even though many of these areas have adequate supplies of the basic nutrients, nitrates and phosphates, they lack essential micronutrients like iron. Ocean fertilization (McElroy 1983) has been widely discussed among scientific specialists, with experiments proposed by the late John Martin (Martin 1990, 1994), and endorsed by the late Professor Roger Revelle, director of the Scripps Oceanographic Institution in La Jolla, California (see Note 2). With the completion and publication of the successful IronEx-II test (see Cooper et al. and other papers in the 10 October 1996 issue of *Nature*), it now makes sense to consider ocean fertilization as a viable candidate for sequestering atmospheric CO_2. The process is cheaper and politically less intrusive than control of emissions (For details, see the Appendix). It also exploits excess CO_2 as a resource for enhancing ocean fisheries.

Additional Benefits: CO_2 as a Resource

The current excess of atmospheric CO_2 can become an important resource, to be exploited for feeding a growing world population. Large-scale fertilization of areas of the Pacific and the Southern Oceans for the purpose of stimulating the growth of phytoplankton would draw down atmospheric CO_2 without depressing the economies of industrialized nations or limiting the economic growth options of developing nations. With phytoplankton as the base of the oceanic food chain, any increase in that population can lead to the development of new commercial fisheries in areas currently devoid of fish. *Carbon dioxide from fossil fuel burning thus becomes a natural resource for humanity rather than an imagined menace to global climate.*

Update on Climate Science

1. The radiative forcing from CO_2 is now about 15% less than previously calculated (Myhre *et al.* 1998). The growth rates of carbon dioxide and methane have slowed markedly in the last few years (Hansen *et al.* 1998). Methane levels have nearly stopped growing (Dlugokencky *et al.* 1998). The cause is not well understood, and there is no assurance that the lower rates will persist.

2. Estimates for the residence time of atmospheric CO_2 continue to decrease. About half will be absorbed into the shallow ocean within thirty years and about 80% within 100 years (Sarmiento, Orr, and Siegenthaler, 1992). A study based on detailed CO_2 data (between 1988 and 1992) has led to the discovery of a huge carbon sink over North America, roughly equal to the amount of the fossil-fuel carbon emitted (Fan *et al.* 1998); the finding has provoked a lively debate.

3. Measurements with high time resolution on the Vostok ice core have established the detailed relationship between CO_2 increases and temperature increases during the transition from an ice age to an in-

terglacial period. During the last three deglaciations, the CO_2 increase was found to lag the temperature increase by about 600 years (Fisher *et al.* 1999). This finding would seem to rule out CO_2 as the cause of the warming. (Methane was ruled out earlier for similar reasons.)

4. The temperature record obtained from weather satellite data is distorted by decay of the satellite orbits (Wentz and Schabel 1998). After applying this correction, as well as a correction for orbit drift, the temperature trend from 1979 to 1997 is not as negative as previously reported, and is close to being zero (Christy *et al.*, 1999). Direct temperature measurements on Greenland ice cores show a cooling between 1940 and 1995 (Dahl-Jensen *et al. 1998),* in support of the satellite data and the (independent) balloon radiosonde data, but not the surface measurements (which show a warming trend that could be due to local contamination, like urban heat islands).

5. None of the data sets, including the surface measurements, agree with the trend currently expected from model results. This trend is calculated to be about 0.25°C per decade at the surface, but rises gradually to twice that value in the upper troposphere (Tett *et al.* 1996).

6. The issue of absorption of solar energy by clouds is still in dispute, as is the feedback of clouds in climate models (Cess *et al* 1990, 1996). The question of the magnitude and sign of the water vapor feedback has not been settled, pending the arrival of better observations (Spencer and Braswell, 1997). A smaller positive or even a negative feedback may explain why temperature trend observations and models disagree.

7. The IPCC report tried to explain the discrepancy between the observed temperature record and model calculations in terms of the cooling effects of manmade sulfate aerosols. This explanation is no longer accepted by many. Tett *et al.* (1996) showed that ozone depletion produces a much stronger impact on tropospheric temperatures than aerosols. Penner *et al.* (1998), exploring the complication of different types of aerosols, concluded that one could not explain the temperature increase of the past century in terms of human influences. Finally, Hansen *et al.* (1998) pointed to the fact that the uncertainty in radiative forcing from aerosols exceeds the uncertainty in the climate sensitivities of the models: "The forcings that drive long-term cli-

mate change are not known with an accuracy sufficient to define future climate change."

8. Increasing numbers of researchers now agree that the warming during the early part of this century is primarily a recovery from the preceding Little Ice Age. If not completely due to solar variability, it must at least contain a strong solar component rather than being manmade (Soon *et al.* 1996; Wigley *et al* 1997).

9. A team of investigators has produced a post-IPCC assessment of tropical cyclones (hurricanes) and their possible relation to global climate change (Henderson-Sellers *et al* 1998). They find substantial multi-decadal variability but no clear evidence for long-term trends. Recent studies indicate that the maximum potential intensity could increase modestly by up to 10 to 20 percent if there is global warming, but these changes are small compared with the observed natural variations. The broad geographical regions affected by tropical cyclones are not expected to change significantly; the popular belief that the region will expand as the oceans warm is a fallacy. The available evidence points to little or no change in expected global frequency.

10. The inverse relation between global warming (or tropical sea surface temperature) and changes in sea level (Singer 1997a) is supported by data on ice accumulation in the Antarctic. It seems increasingly likely that a warming will increase precipitation and ice accumulation, and thus slow down or even reverse the ongoing sea level rise.

Notes

1. A relatively low-cost policy, however, would be to use as much lumber as possible in all permanent structures and reseed existing forests. The carbon in the wood so-used can be expected to remain in storage for 50 years or more, and its harvesting will spur fixation of additional carbon in the new trees grown to replace the old ones—without the need for additional acreage in forests.

2. A personal note: I first heard the idea of ocean fertilization presented by Roger Revelle at the 1990 AAAS meetings in New Orleans. Roger and I discussed this and other approaches to mitigation over breakfast the following morning. The result was an article on the greenhouse warming problem and what to do about it, co-authored by us and Chauncey Starr (Singer et al. 1991).

3. Ice has a very high reflecting power (albedo) in the visible region of the spectrum, while ocean water reflects somewhat less than 10 percent of the incoming solar radiation. Once the oceans freeze over, they could never recover. One of the mysteries of the Earth's early history is how the oceans managed to stay liquid at a time when the solar radiation was less than 80 percent of its present value (Caldeira and Kasting 1992).

4. The clearing of forests and biomass burning have been important CO_2 contributors in the past. Warming of the ocean releases carbon dioxide to the atmosphere. On the other hand, an increase in atmospheric CO_2 stimulates plant growth on land and in the ocean, and thereby decreases the fraction of fossil-fuel CO_2 retained by the atmosphere.

5. A good example is Stephen Schneider's book *The Genesis Strategy: Climate and Global Survival* (1976). Others, like Prof. Reid Bryson of the University of Wisconsin and the late J. Murray Mitchell, NOAA climatologist, in *Global Effects of Environmental Pollution* (1970), ascribed the cooling trend to increasing amounts of air pollution and the increased albedo produced by aerosols, mainly from sulfur oxides emitted in coal burning (see Fig. 14). Their work was ignored during the 1980s as temperatures rose. But aerosols

76

were "rediscovered" after the publication of the first IPCC assessments in 1990 and 1992. By 1995, as climate model results and observations were seen to diverge markedly, aerosols were reintroduced as a means for explaining the discrepancy and for "saving" the large climate sensitivities calculated by conventional general circulation models (GCMs).

6. Hurrell and van Loon (1994) indicate a sudden change in southern hemisphere (SH) circulation, leading also to a strengthening of the Antarctic vortex. Komhyr et al. (1991) remarked on the change in sea surface temperature (SST). It is hardly a coincidence that the Antarctic ozone hole began to grow after 1975, stabilizing around 1987.

7. These conclusions are also based on the assumption that models provide a valid estimate of *natural* climate variability (see Box 2). This ignores such major climate fluctuations as the Little Ice Age, the Medieval Climate Optimum, and other recorded climate changes during the Holocene (Keigwin 1996), the present interglacial that started about 11,000 years ago; none of these climate variations can be accounted for by current GCMs.

8. While nonfossil-fuel energy sources are certainly available, they are not economical at present or face other problems. Biomass, in the form of wood and other materials, has long been an energy source, but is not likely to lead to the generation of concentrated forms of energy required for electric power stations. Solar photovoltaic, wind energy, and geothermal energy are feasible in many locations, but usually require additional and costly storage systems. Hydroelectric and nuclear reactors are proven sources for electric power, but their expansion raises other problems.

9. Azar and Rodhe (1996) have suggested that the present level of CO_2, of the order of 350 ppm, is already "dangerous." Their argument is a circular one; in fact, one cannot define scientifically what constitutes a "dangerous" level (see also note 18).

10. Signatures for the February 27, 1992 Statement are listed on page 92.

11. This statement is based on the International Symposium on the Greenhouse Controversy, held in Leipzig, Germany on November 9-10, 1995, under the sponsorship of the Prime Minister of the State of Saxony. For further information, contact the Europaeische Akademie fuer Umweltfragen (fax +49-7071-72939) or The Science and Environmental Policy Project in Fairfax, Virginia (fax 1-703-352-7535). See also the website www.sepp.org .

12. A similar UHI effect has also been reported from Australia and South Africa (Hughes and Balling 1996, Fig. 18).

13. In fact, aerosols provide a means of testing models by examining hemispheric differences, predicted versus observed.

14. All of the above studies have used global mean data in their comparisons. Since man-induced sulfate is restricted predominantly (~90 percent) to the northern hemisphere, comparisons of observed and predicted changes by hemisphere are more meaningful. Recent reanalyses of the land (the most credible) observations of the southern hemisphere have reduced the estimated warming of the southern hemisphere to about half that of the northern hemisphere (Jones 1994; Hughes and Balling 1996). This appears to represent a serious discrepancy in the argument that sulfates are reducing greenhouse warming.

15. R.G. Johnson (1997) has revived the discussion on Mediterranean outflow affecting the Atlantic circulation, arguing that the construction of the Aswan Dam could lead to glaciation of Labrador. He suggests the construction of a dam at Gibraltar to control outflow.

16. "Lifetime" is related to the fraction of emitted CO_2 retained in the atmosphere. There is no single lifetime since there are different physical processes that remove CO_2—from rapid absorption through surface layers of the oceans, to a medium-term absorption into expanding forests stimulated by CO_2 fertilization, to long-term absorption into the deep ocean mediated by the rate of downwelling. IPCC WGI (1996, p. 76) states, "an approximate value of about 100 years may be given for the adjustment time of CO_2 in the atmosphere, the actual adjustment is faster in the beginning and slower later on."

17. CO_2 levels declined quickly to about present values as a result of the development of vascular plants and weathering, rose to about five times present value about 200 My ago and then declined more or less steadily ever since. (Values much below 200 ppm could seriously impair plant growth.)

18. There is reason to believe that higher CO_2 levels may lead to a more stable climate. Stager and Mayewski (1997) found that climate variability in the most recent ice age, where CO_2 levels were less that 200 ppm, to be greater than in the Holocene, where CO_2 levels ranged around 280 ppm.

Appendix

Mitigation of Climate Change: A Scientific and Economic Appraisal

The global climate doesn't seem to be warming, causing considerable embarrassment to scientists who are banking their reputations on computer models that show large warming trends. And if it did warm, the overall benefits would be positive, causing some embarrassment to economists whose cost-benefit analyses automatically assume (negative) disbenefits. Politicians manage to avoid embarrassment by ignoring both science and economics. They preach policies of mitigation that range from technologically trivial to fantastic, from economically neutral to extremely damaging. Here we appraise a selected group of these policies.

Three Broad Policy Options

The absence of any observed warming—and of a defined goal for the FCCC—certainly creates a problem for politicians; they must persuade their citizens to make major sacrifices to meet a nonexistent threat that cannot be demonstrated but exists only on computer printouts. They are asked to weigh speculative damages of a possible warming against the certain costs of emission controls.

Nevertheless, even with the demonstrated absence of current warming, and with likely benefits should warming occur, we face a situation in which atmospheric CO_2 levels are increasing at a rate of 0.4 percent per year, and methane about twice as fast. We therefore need to tell politicians and the public what can be done to mitigate any climate warming or reduce the growth of greenhouse gases—should that be desired.

There are three general sets of policies that can be followed; we shall discuss them in turn:

1. *Cooling the climate*
2. *Reducing emissions*
3. *Sequestering atmospheric CO_2*

1. Cooling the climate: A large number of schemes have already been proposed for artificially changing the climate (National Academy of Sciences 1991). While all of these are physically possible, some of them are more speculative and costly—and certainly far in the future. They fall into three categories: a) increasing the reflecting power (albedo) of the Earth's surface or atmosphere; b) reducing the incident solar energy flux; and c) modifying the circulation of the atmosphere or ocean.

a. Increasing surface albedo does not appear to be a promising line to follow. Seventy percent of the Earth's surface is covered by oceans which have an albedo of only 9 percent. Floating large white plastic platforms would certainly increase the albedo, but does not make economic or political sense; it would be construed as pollution— and rightly so. Similar objections could be made to increasing atmospheric aerosols or releasing reflective particles into the stratosphere. These processes, of course, occur naturally when volcanoes deposit particulates into the atmosphere, or when sulfur-containing fuels are burned, creating sulfate aerosols.

b. The only method that can claim to be ecologically sound is to launch light-reflecting or absorbing surfaces into space to reduce the incident solar radiation. When combined with the concept of converting solar energy into electricity, such structures, though costly, may make economic sense. Many studies have been published on solar power supplies in space, but the economics has never been persuasive. When the benefits of climate control are added, however, the economics may improve.

c. Affecting the atmospheric (or oceanic) circulation is both difficult and uncertain because of our lack of understanding of all the consequences; but the subject has always held great fascination for geo-scientists. It has been recognized that there are sensitive "pressure points" where ocean circulation might be affected. A recent paper discusses how diversion of fresh water through construction of the Aswan Dam would increase the salinity of the Mediterranean; its outflow through the Straits of Gibraltar would then affect North Atlantic circulation and climate (Johnson 1997).

Climate could be changed through a modification of atmospheric circulation, affecting either cloudiness or the distribution of water vapor, especially its vertical distribution. One method of affecting either one, or both, of these quantities might be through changes in the stratospheric ozone layer. Recent studies (Haigh 1996) suggest that ozone layer changes can affect atmospheric circulation. And there are a variety of ways whereby one can change the total amount of ozone or its horizontal or vertical distribution. Everyone is familiar with methods for destroying ozone in different layers of the stratosphere; but the solar mirrors, referred to above, can also be used to enhance the solar ultraviolet radiation hitting the Earth, thereby creating more ozone in the upper atmosphere.

To sum up: All sorts of speculative schemes have been suggested and will probably continue to appear. Many of them are worthy of research support; a few of them may even justify pilot experiments.

2. Reducing emissions: There are, again, three general categories for reducing emissions.

a. The most benign method, certainly, is energy conservation and a more efficient use of energy through improved capital equipment or processes. In principle, conservation and energy efficiency save not only energy but also money; over time, this has been the main impetus behind such improvements. Problems arise only when efficiency increases are forced through arbitrary standards. A classic example is automobile fuel efficiency—which has increased up to a point because of a public demand for better gasoline mileage. But the demand for roominess and power has also increased the use of pickup trucks and sports vehicles that have very poor fuel consumption.

It is not generally recognized that conservation can be carried too far. Overconservation, which insists on replacing existing capital stock with more energy-efficient equipment, wastes energy—just like underconservation. It leads to the abandonment of equipment that is energy-imbedded and replaces it with equipment that requires energy to construct. As a general rule of thumb, one should not abandon equipment unless the energy savings from replacing it allow a pay-back in less than 3 to 5 years. If the payback period is too long, then energy is being wasted.

b. A different approach to reducing emissions of carbon dioxide is to change to nonfossil-fuel sources of energy: hydroelectric, nuclear, or various kinds of solar-derived sources. Hydroelectric power sources are well established, but require much energy to build, are very site-specific, and cannot be expanded indefinitely as demand grows. Furthermore, they have led to various ecological problems, particularly with fisheries. As a result, the Federal Energy Regulatory Commission is now engaged in tearing down some small privately-owned hydroelectric projects.

Nuclear energy, of course, has problems of its own; they are mainly political, not technical, and might diminish with a good public education program. Nuclear energy is generally safer, less polluting, and often cheaper than electric power systems based on fossil fuels. Since the cost of nuclear electricity is mainly in the capital cost, rather than in the cost of the uranium fuel, economies in construction are particularly important. For example, in France, where plant design has been standardized and where the construction time has been compressed into five years or less, nuclear power is indeed very economic. In the United States, litigation has often extended construction times to more than a decade, incurring huge interest costs. In addition, the Nuclear Regulatory Commission has frequently required post-facto construction changes based on insubstantial safety considerations, further raising the cost (Cohen 1984). The current concerns of spent-fuel disposal and decommissioning of plants add very little to the cost, but figure highly in the public debate.

By contrast, solar energy is everyone's favorite; it seems to have few detractors. But its high capital cost makes it uneconomic, except in specialized applications. A special problem is the storage of energy, needed to supply power at night or on cloudy days. If there exists a large electric grid system, it can furnish storage for a certain amount of solar photovoltaic and wind energy, thus eliminating this large cost factor. It is not likely, however, that any of the solar power systems can become a major supply source for electricity in the near future.

c. The third method of limiting emissions seems to be preferred by regulators. It simply tries to reduce the amount of fossil-fuel burning by rationing or by energy taxes. There are many schemes being discussed, including a hybrid scheme that includes both rationing and a form of taxation. There are difficulties with all of these schemes in deciding upon an equitable distribution of energy, in monitoring its use, and in enforcing limits on fuel burning. These problems are all present for rationing, but are only partially relieved when demand is limited by taxation.

A currently favored scheme would assign an emission quota to each nation—clearly a euphemism for rationing—but permit the buying and selling of unused emission permits, a kind of legalized black market. The problem, of course, arises with the initial allocation of national quotas. Should they be based on present energy consumption, or on the 1990 level, or on some hypothetical future level extrapolated from population growth? And should the per-capita consumption of developing nations be set at some higher level than the present one—and if so, which level? The process is clearly political and may do little to cut total global emissions; more likely, it will create a permanent entitlement program which funnels money from industrialized nations needing emission permits to developing nations willing to sell. It may even have the perverse effect of keeping developing nations from developing, if their governments decide that the transferred funds can be put to a "better" use, like building showy luxury projects or diverting it into foreign bank accounts. Even if the money is not squandered or misappropriated, it is likely to nurture a huge bureaucracy that could seriously throttle free enterprise and economic development in those nations.

Of course, emissions trading is really a hidden energy tax for industrialized nations. Whoever buys the emission permits, whether electric power companies or oil firms, they will have to pass the cost along to the consumer. And if the Protocol resulting from the Climate Treaty aims to stabilize the 1990 atmospheric concentration levels, then CO_2 emissions *worldwide* would have to be cut by more than 60 percent. Even just keeping emissions at the 1990 level requires a carbon tax of $100 a ton or more, and would lead to corresponding increases in energy prices. But these price increases may not be enough to suppress demand. After all, raising gasoline prices by 26 cents a gallon will hardly reduce the demand for driving. For the average motorist, fuel cost is a small fraction of the total cost of automobile transportation—of the order of 20 percent. The price of gasoline would have to go up by several dollars to make a real impact on driving habits.

Worst of all, if emissions were to be limited to 1990 values—or even to values 10 to 20 percent lower—atmospheric concentrations will still increase—albeit somewhat more slowly.

To sum up: Controlling emissions by whatever method is extremely costly, distorts economic decisions, destroys jobs, is difficult to monitor, and practically impossible to enforce. It is likely to create huge international bureaucracies and police forces, damaging not only industrialized countries, but certainly energy exporters and most of the developing countries, since they depend on trade with the industrialized nations (Montgomery 1997). And it would do little good unless emissions worldwide are cut drastically—not just by 10 to 20 percent.

3. Sequestering atmospheric CO_2: Removing CO_2 from the atmosphere can be done by physical methods or biologically. The former have been studied by engineering firms and are judged to be uneconomic under current conditions. Biological sequestration can be done by land-based plants or by biota in the ocean.

The best-studied scheme involves setting up giant forest plantations that can extract CO_2 from the atmosphere. The process is straightforward in that one has to select fast-growing tree species, and find locations where land costs and labor costs are reasonable. A good compilation of the current state of knowledge has been presented by IPCC Working Groups II and III (1996). Unfortunately, quoted cost estimates vary widely, and rise sharply as suitable land becomes scarcer. This is likely to happen because the areas involved are truly very large. If one uses as a rough guide 1 ton of carbon sequestered per hectare per year (Nordhaus 1991), absorbing current emissions would require 50 million square kilometers (ca. 4,500 x 4,500 miles!) Although some attempts have been made by individual firms to plant forests which are said to offset their CO_2 emissions, forest-based sequestration of atmospheric CO_2 has not been pursued on a large scale.

A technique analogous to afforestation, but economically more attractive, is to speed up the natural absorption of CO_2 into the ocean. Currently, much of the world's oceans is a biological desert. Even though many of these areas have adequate supplies of the basic nutrients, nitrates and phosphates, they lack essential micronutrients like iron. Ocean fertilization (McElroy 1983) has been widely discussed among scientific specialists, with experiments proposed by the late John Martin (Martin 1994), and endorsed by the late Professor Roger Revelle, director of the Scripps Oceanographic Institution in La Jolla, California (Singer, Revelle, Starr 1991). With the completion and

---

Mitigation of Atmospheric CO_2: Cost Comparisons

By control of emissions:

Typical cost ~ $100-200 per ton of C emitted[a]

By reforestation:

Typical cost ~ $50 per ton of atmospheric CO_2[b]

equivalent to $25 per ton of CO_2 emitted

By ocean fertilization:[c]

Assume 10^6 tons of Fe (costing ~ $1 billion) can sequester 2×10^{10} tons of C into biomass and needs to be applied yearly. Assume that only 1 percent of C comes from the atmosphere, equivalent to reducing emissions by 2 percent, or 4×10^8 tons.

Therefore, estimated cost ~ $10^9/4 \times 10^8$ ~ $2.50 per ton of C

---

a. Emission control costs are known to rise rapidly as the degree of control increases (IPCC WGIII 1996, Fig. 7.3, p. 254). Besides, at the very best, emission control will only slow down somewhat the rate of increase of concentration.

b. Estimated costs for sequestering CO_2 vary widely, from $7 per ton C (Sedjo and Solomon 1989) to $42–$114 (Nordhaus 1991); see discussion in IPCC WGIII 1996 (pp. 352–355). The cost would rise as the degree of sequestration increases. (Removal of one ton per year is equivalent to reducing emissions by about 2 tons per year.) No allowance has been made for the economic value of the lumber harvested.

c. Comparing the costs in the Table, it is clear that ocean fertilization has by far the lowest cost—$2.50 per ton of carbon—even if one assumes that the phytoplankton draws only 1 percent of carbon from the atmosphere and 99 percent from the oceans. In addition, sequestration by ocean fertilization should exhibit a linear cost curve, unlike forest sequestration or emission control. No allowance is made for the economic value of the fish resources.

publication of the successful IronEx-II test (see research papers in the Oct. 10, 1996 issue of Nature), it now makes sense to consider ocean fertilization as a viable candidate for sequestering atmospheric CO_2. The biomass of phytoplankton in the world's oceans amounts to only one to two percent of the total global plant carbon; yet these organisms fix between 30 to 50 gigatons of carbon (GtC) annually, which is about 40 percent of the total fixed by all biota. (For reference, the atmosphere now contains 750 GtC in the form of CO_2.)

The ocean fertilization experiments, specifically the IronEx-II test, show that, in the equatorial Pacific Ocean at least, the growth of phytoplankton can be dramatically increased by the addition of minute quantities of inorganic iron to surface water. In common with the Southern Ocean, and, to a lesser extent, parts of the northeast Pacific, these waters are termed "high-nutrient, low-chlorophyll" (HNLC), meaning that the normal nutrients are found at the surface, but are not used by the plankton. Addition of the micronutrient iron permits uptake of these unused nutrients and an associated amount of inorganic carbon.

An uncontrolled experiment, the eruption of the volcano Pinatubo, provided an additional test and leads to estimates that can be used for planning a drawdown of atmospheric CO_2. The eruption injected crustal material, about 3 percent iron by weight, into the troposphere and lower stratosphere, and spread it over the globe. Smaller particles may have been carried far enough to enhance productivity in distant HNLC regions, by far the largest of which is the Southern Ocean. Using estimates of the mass deposition flux there, Andrew Watson (1997) figures that the iron deposited amounted to roughly 40,000 tonnes. (This amount is 100,000 times that used in the IronEx-II experiment.) Given a typical carbon/iron molar ratio of 105 for phytoplankton in iron-limited regions, this would enable additional new production, using up about 7×10^{13} mol of carbon. Such an increase would then release a pulse of the order of 10^{14} mol of oxygen into the atmosphere—which is consistent with changes in the hemispheric gradient of the O_2/N_2 ratio observed by R.F. Keeling et al. (1996).

A simple calculation shows that a full-scale demonstration releasing 1 million tons (Mt) of iron in HNLC regions can tie up 20 GtC, which would then be replenished from the atmosphere over some period of time. The drawdown of atmospheric CO_2 would depend on the rate of grazing by zooplankton and higher animals, i.e., on the effectiveness of the "biological CO_2 pump," which rapidly transfers carbon from surface waters to the ocean bottom. There was a slow-down observed in the rate of increase of atmospheric carbon dioxide following the Pinatubo eruption (Sarmiento 1993). It is likely, there-fore, that the atmospheric effect of the proposed demonstration would be measurable by existing CO_2 monitors.

Carrying out the operation would be relatively simple. Single-hulled supertankers exist in surplus; they are not suitable for carry-ing oil cargos but would be ideal for transporting ferrous sulfate, a waste product, and dispersing it—all at low cost. Special patented for-mulations that slow the release of the iron would raise costs some-what but greatly increase the efficiency of the iron absorption. A large-scale demonstration is essential, building on the scientific success of IronEx-II. It would prove the technical and economic feasibility of low-ering the atmospheric CO_2 content at a fraction of the cost now con-templated for emissions reduction. *While it may never be necessary to reduce atmospheric CO_2, it will be comforting to know that we have the technical capability to do so.*

Economic Impact of CO_2 Emission Controls: A Summary of Studies

Cost-benefit analysis is generally subject to great uncertainties; this is even more true in the global warming case. Here, it is not even sure that there are overall disbenefits. Although some individuals, economic sectors, and regions may experience damaging effects, most others should receive positive benefits (see Box 4). Even if one sets aside any scientific doubts and stipulates in advance to an increase in mean temperature, and in its regional distribution, one needs to un-derstand also the overall impact of climate changes on, for example, agricultural production, sea level rise, health effects, etc. The magni-tude and even the direction of these climate impacts are generally disputed; just as for warming itself, there is as yet no scientific con-sensus.

While benefits of warming are in dispute, there is less disagreement about the costs of GH gas controls. The emission goals being considered in the Berlin Mandate process would require a massive restructuring of energy use in the United States. According to the latest data and projections by the U.S. Energy Information Administration, U.S. carbon emissions in 2010 will be about *55 percent greater* than the 1990 levels called for in the Treaty. Despite the significant voluntary efforts as part of the U.S. Climate Change Action Plan, U.S. carbon emissions will increase because of projected economic growth and a growing population.

Impacts on the U.S.

The policies needed to achieve CO_2 emission reductions are as yet unidentified by international negotiators, but surely would have to be quite drastic. A study by the economic consulting firms of Charles River Associates (CRA) and DRI/McGraw-Hill estimates that a tax on the carbon content of fossil fuels in excess of $200 per metric ton likely would be needed in the United States to achieve the goals under discussion. This is equivalent to a new tax of about 60 cents per gallon on gasoline. The prices of most fuels used by residential and commercial customers would increase 50 percent or more with a $200 carbon tax. If taxes were not used to achieve such goals, other policies at least as disruptive would have to be implemented. (CRA/DRI 1994)

I. The economic impacts could be severe. According to the CRA/DRI study, a $100 carbon tax (less than half that necessary to reach the stated goals) could reduce U.S. Gross Domestic Product by 2.3 percent by the year 2100. This is about $203 billion for the economy as a whole, or about $2,000 cost for every U.S. household. The impact also would reduce business fixed investment 4.6 percent, reduce residential invest-ment by 3.2 percent, reduce real consumer spending by $454 per adult, re-duce employment by 500,000, increase inflation, and increase interest rates.

II. A 1997 study (unpublished) for the U.S. Department of Energy by the Argonne National Laboratory analyzed the effect of CO_2 emission controls on six U.S. energy-intensive industries, including chemicals, petroleum refining, paper and allied products, iron and steel, aluminum, and cement. The study simulated the effects of climate change policy through increased fuel prices, based on the carbon content of the various fuels. The results of the study are as follows:

• Climate change actions that increase relative costs to industrialized countries would accelerate the already occurring movement of capital to the developing countries.

• Greenhouse gas emissions would not be reduced significantly; the main effect would be to redistribute output, employment, and GHG emissions from participating to non-participating countries.

The Argonne study details the devastating impact and loss of employment in these energy-consuming industry sectors. Since the results of the study question current U.S. government policy, the study has not been officially accepted.

III. An up-to-date comparison of various cost estimates is given by Christopher Douglas and Murray Weidenbaum (1996). The numbers cited are not simply transfer payments, but real economic costs brought about by a distortion to the economy:

• Stabilizing emissions at 1990 levels, as proposed by the Framework Convention, would generate a net discounted cost of $7 trillion, according to William Nordhaus, Yale University.

• A carbon tax of $100 per ton, which would lower emission levels to near 1990 levels, would cost $203 billion each year according to Lawrence Horwitz of the economic consulting firm DRI/McGraw Hill. Horwitz also projects that 520,000 jobs would be lost each year from 1995 to 2010. A $200 per ton carbon tax would reduce annual GDP by 4.2 percent or $350 billion dollars and result in even greater job losses.

• A DRI/McGraw Hill (1992) study for the U.S. Department of Commerce gives the following results for CO_2 stabilization in the year 2000: a GDP loss of 1.4 percent and tax of $130 per ton of carbon; for a 20 percent CO_2 reduction in the year 2020: a 3 percent GDP loss and carbon tax of $800 per ton C.

IV. Resources Data International (1997) has studied the effect of CO_2 control on the U.S. electricity sector and on GDP, using the current ratio of 1.34 percent growth in electricity for each 1 percent growth in GDP.

• Reducing CO_2 to 1990 levels will limit annual growth rate in the supply of electricity between 1995 to 2015 to 0.83 percent, down from 1.45 percent, under the Department of Energy's (DOE) projected business-as-usual (BAU) scenario. Neither natural gas nor CO_2-neutral generating resources will be able to offset this supply restriction.

- Therefore, up to $1.3 trillion, or 14 percent of GDP, will be at risk in 2010, and up to $16.8 trillion cumulatively from 2005 to 2015.

The study concludes that the proposed emissions-trading program will not work and that U.S. efforts to reduce CO_2 will have diminishing returns:

- The U.S. emitted 23 percent of global CO_2 in 1995, but is projected to emit only 19 percent by 2015 under a BAU scenario, while China and other non-OECD Asian nations also emitted 23 percent of global CO_2 in 1995, but are projected to emit 33 percent by 2015.

- Since only Annex-I nations will be required to control CO_2 emissions under the Berlin Mandate, OECD must reduce its carbon emission by 0.9 Gigatons by 2015 in order to meet 1990 levels; but non-OECD nations will still increase emissions by 2.36 Gt.

Impacts on Other Countries

The impact on Non-Annex I countries has been studied in more detail by David Montgomery of Charles River Associates (1997). Based on the International Impact Assessment Model, developed by Prof. Thomas Rutherford, he reports substantial GDP losses, between 2 and 2.5 percent, by energy exporters and an average loss of 0.5 percent of GDP by other developing countries. The impacts are distributed quite unevenly, with gains by South Korea and India and little change for China. There are also wide variations for the different G-7 countries, with losses ranging from a high of 4.5 percent for Canada to a low of about 1.5 percent for Great Britain (where coal subsides would be removed) and Germany (where the inefficient East Germany economy would be revamped).

The impact of emission controls on fuel-exporting countries has been studied in detail by I.A.H. Ismail of the OPEC secretariat (1997). He assumes uniform carbon taxes to be applied in 1997, rising up to $300 per ton C by 2010. He finds that coal use is affected most, but that world oil demand would drop and so would OPEC production of oil, whence comes the eventual incremental barrel of oil. The major financial losses would be borne by OPEC nations where oil exports form a substantial percentage (up to 50 percent) of GDP. Norway would also be affected since its exports account for 13 percent of GDP.

Conclusions

Policies to limit CO_2 emissions by energy or carbon taxes, while superficially attractive, are economically damaging to the great majority of countries—not only to industrialized nations and energy fuel exporters, but also to the majority of developing nations that use little energy. Such taxes would distort the economy, lower economic growth, raise consumer prices, lower standards of living, and destroy jobs.

In view of the fact that warming predictions are dubious, that warming may result in overall benefits rather than damages, and that the goal of the Climate Treaty cannot as yet be scientifically defined, it seems foolhardy to embark on crash policies that spell economic disaster.

Signatures for the February 27, 1992 Statement

David G. Aubrey, Ph.D., Senior Scientist, Woods Hole Oceanographic Institute; Nathaniel B. Guttman, Ph.D., Research Physical Scientist, National Climatic Data Center; Hugh W. Ellsaesser, Ph.D., Meteorologist, Lawrence Livermore National Laboratory; Richard Lindzen, Ph.D., Center for Meteorology and Physical Meteorology, M.I.T.; Robert C. Balling, Ph.D., Director, Laboratory of Climatology, Arizona State University; Patrick Michaels, Ph.D., Assoc. Professor of Environmental Sciences, University of Virginia; Roger Pielke, Ph.D., Professor of Atmospheric Science, Colorado State University; Michael Garstang, Ph.D., Professor of Meteorology, University of Virginia; Sherwood B. Idso, Ph.D., Research Physicist, U.S. Water Conservation Laboratory; Lev S. Gandin, Ph.D., Visiting Scientist, National Center for Atmospheric Research; John A. McGinley, Chief, Forecast Research Group, Forecast Systems Laboratory, NOAA; H. Jean Thiébaux, Ph.D., Research Scientist, National Meteorological Center, National Weather Service, NOAA; Kenneth V. Beard, Ph.D., Professor of Atmospheric Physics, University of Illinois; Paul W. Mielke, Jr., Ph.D., Professor, Dept. of Statistics, Colorado State University; Thomas Lockhart, Meteorologist, Meteorological Standards Institute; Peter F. Giddings, Meteorologist, Weather Service Director; Hazen A. Bedke, Meteorologist, Former Regional Director, National Weather Service; Gabriel T. Csanady, Ph.D., Eminent Professor, Old Dominion University; Roy Leep, Executive Weather Director, Gillett Weather Data Services; Terrance J. Clark, Meteorologist, U.S. Air Force; Neil L. Frank, Ph.D., Meteorologist; Bruce A. Boe, Ph.D., Director, North Dakota Atmospheric Resource Board; Andrew Detwiler, Ph.D., Assoc. Professor, Institute of Atmospheric Sciences, South Dakota School of Mines and Technology; Robert M. Cunningham, Consulting Meteorologist, Fellow, American Meteorological Society; Steven R. Hanna, Ph.D., Sigma Research Corporation; Elliot Abrams, Meteorologist, Senior Vice President, AccuWeather, Inc.; William E. Reifsnyder, Ph.D., Consulting Meteorologist, Professor Emeritus, Forest Meteorology, Yale University; David W. Reynolds, Research Meteorologist; Jerry A. Williams, Meteorologist, President, Oceanroutes, Inc.; Lee W. Eddington, Meteorologist, Geophysics Division, Pacific Missile Test Center; Werner A. Baum, Ph.D., former Dean, College of Arts & Sciences, Florida State University; David P. Rogers, Ph.D., Assoc. Professor of Research Oceanography, Scripps Institution of Oceanography; Brian Fiedler, Ph.D., Asst. Professor of Meteorology, School of Meteorology, University of Oklahoma; Edward A. Brandes, Meteorologist; Melvin Shapiro, Wave Propagation Laboratory, NOAA; Joseph Zabransky, Jr., Associate Professor of Meteorology, Plymouth State College; James A. Moore, Project Manager, Research Applications Program, National Center for Atmospheric Research; Daniel J. McNaughton, ENSR Consulting and Engineering; Brian Sussman, Meteorologist; Robert D. Elliott, Meteorologist, Fellow, American Meteorological Society; Edward A. Brandes, Meteorologist; Robert E. Zabrecky, Chief Meteorologist; William M. Porch, Ph.D., Atmospheric Physicist, Los Alamos National Laboratory; Earle R. Williams, Ph.D, Associate Professor of Meteorology, Dept. of Earth, Atmospheric, and Planetary Sciences, Massachusetts Institute of Technology; S. Fred Singer, Ph.D., Atmospheric Physicist, University of Virginia, Director, Science & Environmental Policy Project; Please note: Affiliations listed are for identification purposes only.

References

Azar, C., and H. Rodhe. 1997. Targets for Stabilization of Atmospheric CO_2. *Science* 276:1818-19.

Baliunas, S., and R. Jastrow. 1990. Evidence for Long-term Brightness Changes of Solar-type Stars. *Nature* 348:520.

Balling, R.C. 1997. European Temperature Record Since 1751: Where's the Greenhouse Signal? *State of the Climate Report,* P.J. Michaels (ed.), New Hope Environmental Services, Inc.

Barnett, T.P., B.D. Santer, P.D. Jones, R.S. Bradley and K.R. Briffa. 1996. Estimates of Low-frequency Natural Variability in Near-surface Air Temperature. *Holocene* 6:255-65.

Bauer, S.J. 1982. Zum Problem der Sonnenaktivität-Wetter und Klima. *Wetter und Leben* 34:221-26.

Berner, R.A. 1997. The Rise of Plants and Their Effect on Weathering and Atmospheric CO_2. *Science* 276:544-45.

Bottomley, M., C.K. Folland, J. Hsiung, R.E. Newell and D.E. Parker. 1990. *Global Ocean Surface Temperature Atlas: "GOSTA".* A joint project of the Met. Office (Bracknell, England) and the Dept. of Earth, Atmos. and Planet. Science at MIT (Cambridge, Massachusetts).

Broecker, W.S. 1996. Glacial Climate in the Tropics. *Science* 272:1902-04.

Caldeira, K., and J.F. Kasting. 1992. Susceptibility of the Early Earth to Irreversible Glaciation Caused by Carbon Dioxide Clouds. *Nature* 359:226-28.

Callendar, G.S. 1938. The Artificial Production of Carbon Dioxide and its Influence on Temperature. *Quarterly Journal of the Royal Meteorological Society* 64:223-40.

Cess, R.D., G.L. Potter, *et al.* 1990. Intercomparison and Interpretation of Climate Feedback Processes in Nineteen Atmospheric General Circulation Models. *Journal of Geophysical Research.* 95:16,601-16,615

———. 1996. Cloud Feedback in Atmospheric General Circulation Models. *Journal of Geophysical Research.* 101:12,791-12,794.

Charles River Associates and DRI/McGraw-Hill. 1994. *Economic Impacts of Carbon Taxes: Overview and Detailed Results.* Prepared for the Electric Power Research Institute.

Christy, J.R., and J.D. Goodridge. 1995. Precision Global Temperatures from Satellites and Urban Warming Effects of Non-satellite Data. *Atmospheric Environment* 29:1957-61.

Christy, J.R., and R.T. McNider. 1994. Satellite Greenhouse Signal. *Nature* 367:325.

Christy, J.R., R.W. Spencer, and W.D. Braswell. Orbital decay and drift revisions for the MSU tropospheric temperature data sets: Little overall change. *Journal of Geophysical Research* 1999 (submitted)**.

Christy, J.R. 1997. Correspondence (14 March).

Clube, S.V.M., F. Hoyle, W.M. Napier and N.C. Wickramasinghe. 1997. Giant Comets, Evolution and Civilization. (in publication)

Cohen, B.L. 1984. Nuclear Power Economics and Prospects. In *Free Market Energy: The Way to Benefit the Consumer*, edited by S.F. Singer. New York: Universe Books. 218-251.

Cooper, D.J., A.J. Watson and P.D. Nightingale. 1996. Large Decrease in Ocean-Surface CO_2 Fugacity in Response to *In Situ* Iron Fertilization. *Nature* 383:511-13.

Dickinson, R.E. 1975. Solar Variability and the Lower Atmosphere. *Bulletin of the American Meteorological Society* 56:1240-48.

Dlugokencky, E.J., K.A. Masarie, P.M. Lang, and P.P. Tans. 1998. Continuing Decline in the Growth Rate of the Atmospheric Methane Burden. *Nature*. 393:447-450.

Douglas, C., and M. Weidenbaum. 1996. *The Quiet Reversal of U.S. Global Climate Change Policy*. St. Louis. MO: Center for the Study of American Business. *Contemporary Issues Series* 83 (November).

DRI/McGraw-Hill. 1992. *Economic Effects of Using Carbon Taxes to Reduce Carbon Dioxide Emissions in Major OECD Countries*. Prepared for the U.S. Department of Commerce.

Ellsaesser, H.W. 1984. The Climate Effect of CO_2: A Different View. *Atmospheric Environment* 18:431-34; *Atmosfera* 3:3-29 (1990).

Ellsaesser, H.W. 1989. Response to Kellogg's Paper. *In Global Climate Change: Human and Natural Influences*, edited by S.F. Singer. New York: Paragon House. 77.

Fan, S., M. Gloor, J. Mahlman, S. Pacala, J. Sarmiento, T. Takahashi, and P. Tans. 1998. A Large Terrestrial Carbon Sink in North America Implied by Atmospheric and Oceanic Carbon Dioxide Data and Models. *Science*. 282: 442-446.

Fischer H., M. Wahlen, H.J. Smith, D. Mastroianni, and B. Deck. 1999. Carbon Dioxide in the Vostok Ice Core. *Science* 283:1712-1714.

Friis-Christensen, E., and K. Lassen. 1991. Length of the Solar Cycle: An Indicator of Solar Activity Closely Associated with Climate. *Science* 254:698-700.

Gerholm, T.R. 1992. In *The Greenhouse Debate Continued*, edited by S.F. Singer. San Francisco, Calif.: ICS Press.

Goklany, I.M. 1992. Adaptation and Climate Change. Paper presented at the annual meeting of the American Association for the Advancement of Science, Chicago (6-11 February).

Goklany, I.M. 1995. Strategies to Enhance Adaptability: Technological Change, Sustainable Growth and Free Trade. *Climatic Change* 30:427-49.

Goldemberg, J. 1995. Energy Needs in Developing Countries and Sustain-ability. *Science* 269:1058-59.

Goodridge, J.D. 1996. Comments on Regional Simulations of Greenhouse Warming Including Natural Variability. *Bulletin of the American Meteorological Society* 77:3-4 (July).

Gore, A. 1995. Speech given at the annual meeting of the American Association for the Advancement of Science, Baltimore, Maryland.

Haigh, J.D. 1996. The Impact of Solar Variability on Climate. *Science* 272:981-84.

Hansell, D.A., N.R. Bates and C.A. Carlson. 1997. Predominance of Vertical Loss of Carbon from Surface Waters of the Equatorial Pacific Ocean. *Nature* 386:59-61.

Hansen, J.E., M. Sato, A. Lacis, R. Ruedy, I. Tegen, and E. Matthews. 1998. Climate Forcings in the Industrial Era. *Proceedings of the National Academy of Sciences USA.* 95: 12753-12758.

Hansen, J., and S. Lebedeff. 1987. Global Trends of Measured Surface Air Temperature. *Journal of Geophysical Research* 92:13345-72.

Hansen, J., H. Wilson, M. Sato, R. Ruedy, K. Shah and E. Hansen. 1995. Satellite and Surface Temperature Data at Odds? *Climatic Change* 30:103-17.

Hansen, J., M. Sato and R. Ruedy. 1997. Radiative Forcing and Climate Response. *Journal of Geophysical Research* 102(D6):6831-64.

Hasselmann, K. 1997. Are We Seeing Global Warming? *Science* 276:914-15.

Henderson-Sellers, A., H. Zhang, G. Berz, K. Emanuel, W. Gray, C. Landsea, G. Holland, J. Lighthill, S-L. Shieh, P. Webster, and K. McGuffie. 1998. Tropical Cyclones and Global Climate Change: A Post-IPCC Assessment. *Bulletin of the American Meteorological Society.* 79: 19-38.

Hughes, W.S. 1992. Greenhouse/Global Warming and Temperature Measurement. Melbourne, Australia: Tasman Institute (March). 2-3.

Hughes, W.S. and R.C. Balling. 1996. Urban Influences on South African Temperature Trends. *International Journal of Climatology* 16:935-40.

Hurrell, J.W., and H. van Loon. 1994. A Modulation of the Atmospheric Annual Cycle in the Southern Hemisphere. *Tellus* 46A:325-38.

Hurrell, J.W., and K.E. Trenberth. 1996. Satellite versus Surface Estimates of Air Temperature since 1979. *Journal of Climate* 9:2222-32.

Hurrell, J.W., and K.E. Trenberth. 1997. Spurious Trends in Satellite MSU Temperatures from Merging Different Satellite Records. *Nature* 386:164-67.

Idso, S.B. 1989. *Carbon Dioxide and Global Change: Earth in Transition.* Tempe, Arizona: IBR Press.

Idso, S.B. 1995. *CO_2 and the Biosphere: The Incredible Legacy of the Industrial Revolution.* St. Paul, Minn.: Department of Soil, Water and Climate, University of Minnesota.

IPPC WGI. 1990. *Climate Change: The IPCC Scientific Assessment,* edited by J.T. Houghton, G.J. Jenkins and J.J. Ephraums. Cambridge, England: Cambridge University Press.

IPCC WGI. 1996. *Climate Change 1995: The Science of Climate Change,* edited by J.T. Houghton, L.G. Meira Filho, B.A. Callander, N. Harris, A. Kattenberg and K. Maskell. Cambridge, England: Cambridge University Press.

IPCC WGII. 1996. *Climate Change 1995: Impacts, Adaptations and Mitigation of Climate Change: Scientific-technical Analyses,* edited by R.T. Watson, M.C. Zinyowera, R.H. Moss and D.J. Dokken. Cambridge, England: Cambridge University Press.

IPCC WGIII. 1996. *Climate Change 1995: Economic and Social Dimensions of Climate Change,* edited by J.P. Bruce, H. Lee and E.F. Haites. Cambridge, England: Cambridge University Press.

Ismail, I.A.H. 1997. *Vulnerability to and Impacts of Climate Change Measures on Oil and Fossil Fuel Producing/Exporting Countries.* Vienna, Austria: OPEC Workshop on the Environment (February).

Jacoby, G.C., R.D. D'Arrigo and T. Davaajamts. 1996. Mongolian Tree Rings and 20th-century Warming. *Science* 273:771-73.

Jain, A.K., H.S. Kheshgi and D.J. Wuebbles. 1996. A Globally Aggregated Reconstruction of Cycles of Carbon and its Isotopes. *Tellus* 48B:583-600.

Johnson, R.G. 1997. Climate Control Requires a Dam at the Strait of Gibraltar. *Eos, Transactions of the American Geophysical Union* 78:227-81.

Jones, P.D. 1994. Hemispheric Surface Air Temperature Variations: A Reanalysis and an Update to 1993. *Journal of Climate* 7:1794-802.

Karl, T.R., and P.D. Jones. 1989. Urban Bias in Area-averaged Surface Air Temperature Trends. *Bulletin of the American Meteorological Society* 70:265-70.

Karl, T.R., G. Kukla, V.N. Razuvayev, M.J. Changrey, R.G. Quayle, R.R. Heim, Jr., D.R. Easterling and C.B. Fu. 1991. Global Warming: Evidence for Asymmetric Diurnal Temperature Change. *Geophysical Research Letters* 18:2253-56.

Keeling, C.D., J. F. S. Chin, and T. P. Whorf. 1996. Increased Activity of Northern Vegetation Inferred from Atmospheric CO_2 Measurements. *Nature* 382:146-49.

Keeling, C.D., R.B. Bacastow, A.F. Carter, S.C. Piper, T.P. Whorf, M. Heiman, W.G. Mook and J. Roeloffzen. 1989. A Three-dimensional Model of Atmospheric CO_2 Transport Based on Observed Winds: I. Analysis of Observational Data. Washington, D.C.: American Geophysical Union. *Geophysical Monograph* 55:165-236.

Keeling, R.F., S.C. Piper and M. Heimann. 1996. Global and Hemispheric CO_2 Sinks Deduced from Changes in Atmospheric CO_2 Concentration. *Nature* 381:218-21.

Keigwin, L.D. 1996. The Little Ice Age and Medieval Warm Period in the Sargasso Sea. *Science* 274:1504-08.

Kerr, R.A. 1994. Climate Modeling's Fudge Factor Comes under Fire. *Science* 265:1528.

Kerr, R.A. 1995. Darker Clouds Promise Brighter Future for Climate Models. *Science* 267:454.

Kerr, R.A. 1997a. A New Driver for the Atlantic's Moods and Europe's Weather? *Science* 275:754-55.

Kerr, R.A. 1997b. Greenhouse Forecasting still Cloudy. *Science* 276:1040-42.

Komhyr, W., et al. 1991. Possible Influence of Long-term Sea-surface Temperature Anomalies in the Tropical Pacific on Global Ozone. *Canadian Journal of Physics* 69:1093-1102.

Labitzke, K., and H. van Loon. 1989, 1988, 1988, 1994. Association Between the 11-year Solar Cycle, the QBO, and the Atmosphere. *Journal of Climate* 2:554-65; *Journal of Atmospheric and Terrestrial Physics* 50:197; *Journal of Climate* 1:905-20; *Meteorological Zeitschrift* 3:259.

Landsea, C.W., N. Nicholls, W.M. Gray and L.A. Avila. 1996. Downward Trends in the Frequency of Intense Atlantic Hurricanes during the Past Five Decades. *Geophysical Research Letters* 23(13):1697-1700.

Lassen, K., and E. Friis-Christensen. 1995. Variability of the Solar Cycle Length during the Past Five Centuries and the Apparent Association with Terrestrial Climate. *Journal of Atmospheric and Terrestrial Physics* 57:835-845.

Lean, J. 1991. Variations in Sun's Radiative Output. *Reviews of Geophysics* 29:505-35.

Lichtblau, J.H. 1998. *Energy Demand and CO_2 Emissions*. New York: Petroleum Industry Research Foundation. An appendix to *Global Warming: Unfinished Business*, by S.F. Singer. Oakland, Calif.: The Independent Institute (in publication).

Linden, H.R. 1998. Are the IPCC Carbon Emission and Carbon Dioxide Stabilization Scenarios Realistic? An appendix to *Global Warming: Unfinished Business*, by S.F. Singer. Oakland, Calif.: The Independent Institute (in publication).

Lindzen, R.S. 1990. Some Coolness Concerning Global Warming. *Bulletin of the
American Meteorological Society* 71:288-99.

Manabe, S., and R.T. Wetherald. 1967. Thermal Equilibrium of the Atmosphere
with a Given Distribution of Relative Humidity. *Journal of Atmospheric Science* 24:241-59.

Manne, A.S., and R.G. Richels. 1991. Global CO_2 Emission Reductions: The Impact of Rising Energy Costs. *Energy Journal* 12:87-107.

Markson, R., and M. Muir. 1980. Solar Wind Control of the Earth's Electric Field.
Science 208:979.

Marland, G. and B. Schlamadinger. 1995. Biomass Fuels and Forest-Management
Strategies: How Do We Calculate the Greenhouse-gas Emissions Benefits?
Energy 20:1131-40.

Marland, G., R.J. Andres and T.A. Boden. 1994. Global, Regional, and National CO_2
Emissions. In *Trends 93: A Compendium of Data on Global Change*, edited
by T.A. Boden, D.P. Kaiser, R.J. Sepanski and F.W. Stoss. Oak Ridge, Tenn.:
Oak Ridge Nat. Laboratory. ORNL/CDIAC-65:505-84.

Martin, J.H., et al. 1994. The Iron Hypothesis: Ecosystem Tests in Equatorial Pacific Waters. *Nature* 371:123-29.

Mason, B.J. 1995. Book Review. *Environmental Conservation* 22:323-24.

McElroy, M.B. 1983. Marine Biological Controls on Atmospheric CO_2 and Climate.
Nature 302:328-29.

Meadows, D.H., D.L. Meadows, J. Randers, and W.W. Behrens III. 1972. *Limits to
Growth*. New York: Universe Books.

Mendelsohn, R. and J. E. Neumann (eds.). The Impact of Climate Change on the
United States Economy. Cambridge University Press, Cambridge: 1999.

Mendelsohn, R., W.D. Nordhaus and D. Shaw. 1994. The Impact of Global Warming on Agriculture: A Ricardian Analysis. *American Economic Review* 84(4):753-
71.

Michaels, P.J. 1995. Looking for Answers. *World Climate Report* 1(6):1-2.

Michaels, P.J. 1996. *State of the Climate Report*. Arlington, VA: Western Fuels Association.

Michaels, P.J. 1997. U.S. Senate Report Misled. *World Climate Report* 2(14):3.

Michaels, P.J., and P.C. Knappenberger. 1996. Human Effect on Global Climate?
Nature 384:522-523.

Mitchell, J.F.B., and T.C. Johns. 1997. On Modification of Global Warming by Sulfate Aerosols. *Journal of Climate* 10:245-66.

Mitchell, J.F.B., T.C. Johns, J.M. Gregory and S.F.B. Tett. 1995. Climate Response
to Increasing Levels of Greenhouse Gases and Sulphate Aerosols. *Nature*
376:501-04.

Mitchell, J.M. Jr., 1970. 1975. A Reassessment of Atmospheric Pollution as a Cause
of Long-term Changes of Global Temperature. In *Global Effects of Environmental Pollution*, edited by S.F. Singer. Dordrecht-Holland and Boston: Reidel
Publ. Co.; ———. *The Changing Global Environment*, edited by S.F. Singer.
Dordrecht-Holland and Boston: Reidel Publ. Co.

Montgomery, W.D. 1997. *Impacts of Annex-I Country Commitments on Non-Annex-I Countries*. Vienna, Austria: OPEC Workshop on the Environment (20 February).

Moore, T.G. 1995. *Global Warming: A Boon to Humans and Other Animals*. Stanford, Calif.: Hoover Institution. Stanford University.

Myhre, G., E.J. Highwood, K.P. Shine, and F. Stordal. 1998. New Estimates of Radiative Forcing Due to Well Mixed Greenhouse Gases. *Geophysical Research Letters*. 25: 2715-2718.

Myneni, R.B., C.D. Keeling, C.J. Tucker, G. Asrar and R.R. Nemani. 1997. Increased Plant Growth in the Northern High Latitudes from 1981 to 1991. *Nature* 386:698-702.

National Academy of Sciences. 1991. *Policy Implications of Greenhouse Warming: Mitigation, Adaptation, and the Science Base*. Washington D.C.: National Academy Press.

Nordhaus, W.D. 1991a. The Cost of Slowing Down Climate Change: A Survey. *Energy Journal* 12(1):37-65.

Nordhaus, W.D. 1991b. To Slow or Not to Slow: Economics of the Greenhouse Effect. *Economic Journal* 101:920-937.

Oerlemans, J. 1982. Response of the Antarctic Ice Sheet to Climate Warming. *Journal of Climate* 2:1-12.

Parry, M., Arnell, N., Hulme, M., Nicholls, R. and M. Livermore. "Adapting to the inevitable." Nature, 395, 741, 1998.

Patz, J.A., P.R. Epstein, T.A. Burke, J.M. Balbus. 1996. Global Climate Change and Emerging Infectious Diseases. *Journal of the American Medical Association* 275:217-23 (17 January); S.F. Singer. 1996. Reply. *Journal of the American Medical Association* 276:373 (7 August).

Penner, J.E., C.C. Chuang, and K. Grant. 1998. Climate Forcing by Carbonaceous and Sulfate Aerosols. *Climate Dynamics*. ***

Rasool, S.I., and S.H. Schneider. 1971. Atmospheric Carbon Dioxide and Aerosols: Effects of Large Increases on Global Climate. *Science* 173:138-41.

Rasool, S.I., and S.H. Schneider. 1972. Aerosol Concentrations: Effect on Planetary Temperatures. *Science* 175:96.

Revelle, R. 1977. In *Arid Zone Development: Potentialities and Problems*, edited by Y. Mundlak and S.F. Singer. Cambridge, Mass.: Ballinger Publishing.

Revelle, R., and H.E. Suess. 1957. Carbon Dioxide Exchange between Atmosphere and Ocean and the Question of an Increase of Atmospheric CO_2 during the Past Decades. *Tellus* 9:18-27.

Santer, B.D., K.E. Taylor, T.M.L. Wigley, J.E. Penner, P.D. Jones and U. Cubasch. 1995. Towards the Detection and Attribution of an Anthropogenic Effect on Climate. *Climate Dynamics* 12:79–100.

Santer, B.D., K.E. Taylor, T.M.L. Wigley, P.D. Jones, D.J. Karoly, J.F.B. Mitchell, A.H. Oort, J.E. Penner, V. Ramaswamy, M.D. Schwarzkopf, R.J. Stouffer and S. Tett. 1996. A Search for Human Influences on the Thermal Structure of the Atmosphere. *Nature* 382:39-46.

Sarmiento, J.L., J.C. Orr, and U. Siegenthaler. 1992. A Perturbation Simulation of CO_2 Uptake in an Ocean General Circulation Model. *Journal of Geophysical Research*. 97: 3621-3646.

Sarmiento, J.L. 1993. Atmospheric CO_2 Stalled. *Nature* 365:697.

Schlesinger, M.E., and X. Jiang. 1991. Revised Projections of Future Greenhouse Warming. *Nature* 350:219-21.

Schneider, S.H. 1976. *The Genesis Strategy: Climate and Global Survival.* New York: Plenum Press.

Schneider, S.H. 1992. Introduction to Climate Modeling. In *Climate System Modeling*, edited by K.E. Trenberth. New York: Cambridge University Press.

Schwartz, S.E. and M.O. Andreae. 1996. Uncertainty in Climate Change Caused by Aerosols. *Science* 272:1121-22.

Sedjo, R.A. and A.M. Solomon. 1989. Climate and Forests. In *Greenhouse Warming: Abatement and Adaption*, edited by N.J. Rosenberg, W.E. Easterling III, P.R. Crosson, and J. Darmstadter. Washington, D.C.: Resources for the Future.

Singer, S.F. 1996. Climate Change and Consensus. *Science* 271:581.

Singer, S.F. 1997a. Global Warming Will Not Raise Sea-levels. Abstract for 1997 Fall Meeting of the American Geophysical Union (submitted for publication).

Singer, S.F. 1997b. Climate Warming from Increasing Air Traffic? Presented at NASA Conference on the Atmospheric Effects of Aviation, Virginia Beach, Virginia (10-14 March).

Singer, S.F. 1997c. A Treaty Built on Hot Air ... Not Scientific Consensus, *The Wall Street Journal* (25 July).

Singer, S.F., R. Revelle and C. Starr. 1991. What to Do about Global Warming: Look before You Leap. *Cosmos* 1(1):28-33.

SMIC. 1971. *Inadvertent Climate Modification: Report of the Study of Man's Impact on Climate*, edited by C.L. Wilson and W.H. Matthews. Cambridge, Mass.: MIT Press.

Soon, W.H., E.S. Posmentier, and S.L. Baliunas. 1996. Inference of Solar Irradiance Variability from Terrestrial Temperature Changes, 1880-1993. *Astrophysical Journal*. 472: 891-902.

Spencer, R.W., and J.R. Christy. 1990. Precise Monitoring of Global Temperature Trends from Satellites. *Science* 247:1558-62.

Spencer, R.W., and J.R. Christy. 1992. Precision and Radiosonde Validation of Satellite Gridpoint Temperature Anomalies, Part II: Tropospheric Retrieval and Trends during 1979-90. *Journal of Climate* 5:858-66.

Spencer, R.W. and W.D. Braswell. 1997. How Dry Is the Tropical Free Troposphere? Implications for Global Warming Theory. *Bulletin of the American Meteorological Society* 78(6):1097-106.

Stager, J.C., and P.A. Mayewski. 1997. Abrupt Early to Mid-Holocene Climatic Transition Registered at the Equator and the Poles. *Science* 276:1834-36.

Sun, D.Z., and R.S. Lindzen. 1993. Distribution of Tropical Water Vapor. *Journal of Atmospheric Sciences* 50:1643-60.

Svensmark, H., and E. Friis-Christensen. 1997. Variation of Cosmic Ray Flux and Global Cloud Coverage: A Missing Link in Solar-climate Relationships. *Journal of Atmospheric and Terrestrial Physics* 59(11):1225.

Taylor, K.E. and J.E. Penner. 1994. Response of the Climate System to Atmospheric Aerosols and Greenhouse Gases. *Nature* 369:734-36.

Tett, S.F.B., J.F.B. Mitchell, D.E. Parker, M.R. Allen. 1996. Human Influence on the Atmospheric Vertical Temperature Structure: Detection and Observations. *Science* 274:1170-73.

Thompson, S.L. and D. Pollard. 1995. Greenland and Antarctic Mass Balances for Present and Doubled Atmospheric CO_2 from the GENESIS Version 2 Global Climate Model. *Eos, Transactions of the American Geophysical Union* 76(46) Supp. (meeting abstracts); A Global Climate Model (GENESIS) with a Land-surface Transfer Scheme (LSX), Part I: Present Climate Simulation. *Journal of Climate* 8(4):732-61.

Tinsley, B.A. 1996. Correlations of Atmospheric Dynamics with Solar Wind Induced Changes of Air-earth Current Density into Cloud Tops. *Journal of Geophysical Research* 101:29701.

Tinsley, B.A. and G.W. Deen. 1991. Apparent Tropospheric Response to MeV-GeV Particle Flux Variations: A Connection via Electrofreezing of Supercooled Water in High-level Clouds. *Journal of Geophysical Research* 96:22283-96.

Trupin, A. and J. Wahr. 1990. Spectroscopic Analysis of Global Tide Gauge Sea Level Data. *Geophysical Journal International* 100:441-53.

U.S. Department of Agriculture. 1952. Insects. In *The Yearbook of Agriculture, 1952.* Washington, D.C.: Government Printing Office. 147-160.

U.S. General Accounting Office. 1995. *Global Warming: Limitations of General Circulation Models and Costs of Modeling Efforts.* Washington D.C.: Government Accounting Office. Document GAO-RCED-95-164.

Watson, A.J. 1997. Volcanic Iron, CO_2, Ocean Productivity and Climate. *Nature* 385:587-88.

Weart, S.R. 1997. The Discovery of the Risk of Global Warming. *Physics Today* 50:34-40.

Wentz, F.J. and M. Schabel. 1998. Effects of Orbital Decay on Satellite-Derived Lower-Tropospheric Temperature Trends. *Nature.* 394: 661-664.

Wigley, T.M.L., and S.C.B. Raper. 1992. Implications on Climate and Sea Level of Revised IPCC Emissions Scenarios. *Nature* 357:293-300.

Wigley, T.M.L., P.D. Jones, and S.C.B. Raper. 1997. The Observed Global Warming Record: What Does It Tell Us?. *Proceedings of the National Academy of Sciences USA.* 94: 8314-8320.

Wigley, T.M.L., R. Richels and J.A. Edmonds. 1996. Economic and Environmental Choices in the Stabilization of Atmospheric CO_2 Concentrations. *Nature* 379:240-43.

Wittwer, S.H. 1995. Flower Power: Rising Carbon Dioxide Is Great for Plants. *Policy Review* (Fall). 4-9.

Wittwer, S.H. 1992. *Food, Climate, and Carbon Dioxide: The Global Environment and World Food Production.* Boca Raton, Florida: CRC Lewis Publishers.

About the Author

S. FRED SINGER is one of the preeminent authorities on energy and environmental issues. A pioneer in the development of rocket and satellite technology, Dr. Singer designed the first satellite instrument for measuring atmospheric ozone and was a principal developer of scientific and weather satellites.

Research Fellow at the Independent Institute in Oakland, California, and President of the Science and Environmental Policy Project, Dr. Singer is Professor Emeritus of Environmental Science and a member of the Energy Policy Studies Center at the University of Virginia; Distinguished Research Professor, Institute for Space Science and Technology; and Distinguished Research Professor at George Mason University. He received his Ph.D. in physics from Princeton University, and he is the recipient of the White House Special Commendation, Gold Medal Award from the U. S. Department of Commerce, and (First) Science Award from the British Interplanetary Society. A fellow of the American Association for the Advancement of Science, Dr. Singer has received an Honorary Doctorate from Ohio State University, and was elected to the International Academy of Astronautics.

Dr. Singer has served as Vice Chairman of the National Advisory Committee on Oceans and Atmospheres; Chief Scientist for the U. S. Department of Transportation; Deputy Assistant Administrator at the U. S. Environmental Protection Agency; Deputy Assistant Secretary at the U. S. Department of the Interior; (First) Dean of the School of Environmental and Planetary Sciences, University of Miami; (First) Director of the U. S. Weather Satellite Center (Department of Commerce); Director of the Center for Atmospheric and Space Physics, University of Maryland; and Research Physicist, Upper Atmospheric Rocket Program, Johns Hopkins University. He has also been a visiting scholar at the Woodrow Wilson International Center for Scholars; Jet Propulsion Laboratory, California Institute of Technology; National Air and Space Museum; Lyndon Baines Johnson School for Public Affairs, University of Texas; and the Soviet Academy of Sciences Institute for Physics of the Earth.

A consultant to numerous government agencies, businesses, and other organizations, Dr. Singer has served on numerous federal advisory panels concerned with space, oceans and the atmosphere, and he has testified before Congress on acid rain, nuclear waste, and global environmental issues.

Dr. Singer is the author or editor of fourteen books including *The Changing Global Environment, Free Market Energy, Global Climate Change, Is There an Optimum Level of Population?, Manned Laboratories in Space, The Ocean in Human Affairs*, and *The Universe and Its Origin*. He is also the author of over 400 technical articles in scientific, economics and public policy journals plus over 200 articles in the *Wall Street Journal, New York Times, Los Angeles Times, Washington Post, Chicago Tribune, Newsweek, New Republic, National Review, Christian Science Monitor, Reader's Digest, Cleveland Plain Dealer* and other popular publications. In addition, Dr. Singer has been featured in articles in *Time, Life*, and *U. S. News & World Report*, and he has been interviewed on *Nightline, Today Show, MacNeil-Lehrer News Hour, Nightwatch*, and many other national television news programs.

Index